**DO NOT REMOVE
CARDS FROM POCKET**

CUT YOUR ELECTRIC BILLS IN HALF

by Ralph J. Herbert, Ph.D.

Illustrations by John Carlance

 Rodale Press, Emmaus, Pennsylvania

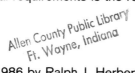

Printed in the United States of America on recycled paper, containing a high percentage of de-inked fiber.

Book design by Anita G. Patterson

Library of Congress Cataloging-in-Publication Data
Herbert, Ralph J., 1943–
    Cut your electric bills in half.

    Includes index.
    1. Dwellings—Energy conservation. 2. Electric utilities—Costs. 3. Electric power—Conservation. I. Title.
TJ163.5.D86H46   1986   644   85–25804
ISBN 0-87857-595-2 hardcover
ISBN 0-87857-596-0 paperback

2 4 6 8 10 9 7 5 3 1   hardcover
2 4 6 8 10 9 7 5 3 1   paperback

This book is dedicated
to my wife, Karen,
to my mother and father,
and to my children,
Ralph, Jennifer, and Jonathan.

Contents

Acknowledgments viii

PART 1

Chapter 1	Overview	1
Chapter 2	Hot Water	12
Chapter 3	Lights	25
Chapter 4	Refrigeration	32
Chapter 5	Cooking	37
Chapter 6	Cooling	44
Chapter 7	Heating	56
Chapter 8	Other Appliances	86
Chapter 9	Conclusion	94

PART 2

Section 1	Energy-Efficient Water Heaters	102
Section 2	Tankless Gas Water Heaters	105
Section 3	Residential Solar Water Heaters	107
Section 4	Windows, Skylights, and Related Devices	110
Section 5	Timing Devices	115
Section 6	Refrigerators and Freezers	117
Section 7	Ranges and Ovens	118
Section 8	Energy-Efficient Air Conditioners	119
Section 9	Fans	122
Section 10	Air-to-Air Heat Pumps	125
Section 11	Energy-Efficient Furnaces	127
Section 12	Passive-Solar Space Heating	130
Section 13	Wood Stoves	133
Section 14	Air-to-Air Heat Exchangers	135
Section 15	Dishwashers, Clothes Washers, and Dryers	136
Section 16	Alternative Generation of Electricity	137

Index 141

Acknowledgments

I would like to thank Karl Grossman for his advice and encouragement, and express my appreciation for the generous editorial assistance of the late Lynne Liot. My thanks also to Millie at Inside Address and to my editor at Rodale Press for his skillful and expeditious editorial contribution to the manuscript. Very special thanks to my wife, Karen Wagner, for her suggestions, encouragement, and typing.

PART 1

Chapter 1
Overview

Electricity is the most expensive form of energy used in our homes, yet for many tasks it is either unnecessary or used wastefully. This book will show you how to reduce your consumption of electrical power and how to save a substantial amount of money without having to sacrifice your present lifestyle.

The cost of residential electricity in the United States ranges from 2¢ per kilowatt-hour (kwh) up to 15¢ per kwh. No matter what rate you are charged, you probably pay more than you should for electricity; almost everyone in America does.

The average cost of a kilowatt-hour of electrical energy is about 9¢, and a typical family uses about 600 kwh of power each month. Thus, many families face electric bills of around $54 a month, not counting sales taxes and other miscellaneous charges. Millions of Americans pay more than $54 per month, however, and during the next few years, unprecedented rate hikes of 25 to 80 percent are anticipated in many sections of the nation.

In northeastern and southwestern states, monthly electric bills of $75 to $90 are already common. Families with "big-draw" appliances like electric ranges, water heaters, baseboard resistance space heaters, and central air conditioners often pay hundreds of dollars each month to the electric company. Rate increases in regions where bills are already high could place an almost unbearable financial burden on low- and middle-income households. In areas where rates have until now been moderate, higher bills would cut heavily into the disposable income of many families. The economic stress generated by the burden of high electric bills has in fact become so widespread that it has been given a name of its own—rate shock.

The impact of rate shock is already visible in several areas of the country. In New York State, for example, the typical customer of the Long Island Lighting Company (LILCO) pays over $1,000 a year for electricity. And some families pay much more. Long Island families with major electrical appliances can pay $2,000 or even $3,000 a year. Many Long Islanders speak of leaving the area

1

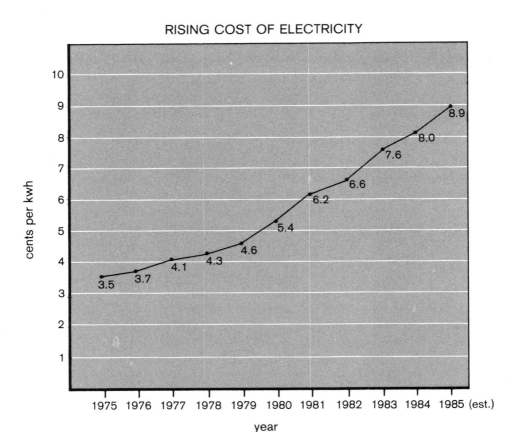

RISING COST OF ELECTRICITY

Figure 1-1. Electric rates have risen rapidly in recent years, and the trend seems destined to continue. (The rates presented here are based on 500-kwh use per month.)

rather than facing the dual burden of high taxes and prohibitive electric bills. Numerous major firms have considered moving elsewhere specifically because of high LILCO rates. But Long Island is not the only part of the nation confronting rate shock. With electrical companies in many other states seeking huge rate hikes in the near future, a great many households face the possibility of equally high electric bills. Residents of San Diego, California, already pay as much as Long Islanders, and Con Edison customers in New York City pay even more. The effects of rate shock will soon become commonplace across America, with billions of additional dollars being drained from the customers of the electric companies.

Several factors underlie the emerging electric rate crisis in America. For several decades prior to the 1970s, the demand for electricity grew at an average

annual rate of about 7 percent. Utilities, expecting that the growth in consumption would continue indefinitely, made commitments to new plant construction. But by the late 1970s, the growth in demand for electricity had fallen off to just 1 to 3 percent a year, and by the early 1980s, it had fallen to 0 percent. As the new power plants—begun years earlier—came on-line, generating capacity climbed much faster than the need for new power. Today the electrical generating capacity of the electric industry is about 50 percent above what we need. Despite this huge excess, new plants continue to come on-line and utilities aggressively pursue the completion of additional power plants in order to get a return on stockholder investments.

Electric companies that have invested heavily in nuclear plant construction face the greatest dilemma. Cost overruns of 1,100 percent are common. Nuclear power, originally viewed to be the cheapest source of electrical power, is now considered the most expensive source. In the past, electric utilities with nuclear-generating capacity exhibited greater rate increases than nonnuclear utilities. In the future, customers of electric companies with newly completed nuclear facilities will face substantial rate increases. The irony is that many of these outrageously expensive power plants will never be needed.

What You Can Do

Ratepayers wring their hands in frustration because they feel there is nothing they can do to combat high electric bills. They are wrong. Although it is true that we need electricity to do various tasks around our homes, we actually require far less electrical power than we currently use. Many people around the country are beginning to realize that there are simple, cheap ways to save on electricity and that in some cases the switch to alternative fuels and new equipment can save them substantial amounts of money over the long run.

The purpose of this book is to show you—the ratepayer—how to simply and inexpensively reduce your electric bill by half. I originally decided to write this book because many friends and neighbors were paying three and four times as much as I was for electricity each month. In my area, the electric rate was over 11¢ per kwh and the average bill was running $72 per month. My bill was $25 to $30 per month for a family of four. People would ask me if I sat in the dark, did I have a TV, perhaps a root cellar instead of a refrigerator? Was I cold in winter?

My quality of life was in fact equal to theirs. I lived in 1,600 square feet of well-lighted space, stayed warm in winter and cool in summer, took a hot shower every day, and had two televisions. I lived as well as my neighbors did, but I used less electricity to maintain my lifestyle. You can also live well while spending less each month for electricity than you do now. Much less.

For a small, once-only investment of about $100, a typical family can reduce its overall electricity consumption by 25 percent without any adverse impact on

its standard of living. A moderate investment of $400 to $500 can bring you a permanent 50 percent reduction in the consumption of household electricity, without hurting your lifestyle. This modest investment will normally be returned to you in savings within two years or less. In many regions of the United States, the payback would come in a year or less. Families with big-draw appliances can save up to 80 percent on the operation of these machines by using the conservation tips I offer in this book. These appliances include space heaters, water heaters, and air conditioners. Savings of 30 percent or more are easily achieved on refrigeration, cooking, lighting, and on the use of a number of smaller appliances—without having to replace existing equipment. Even greater reductions—50 percent to 60 percent—are possible in these areas over the long run by simply replacing existing equipment as it wears out with newer high-efficiency models. The information I give you in this book should not only enable you to avoid the effects of rate shock, but it may also ensure that years from now you will pay no more for electricity each month than you do today.

This book is organized according to various categories of electrical use—lighting, cooling, cooking, etc. In these various categories, I will describe conservation measures that include both simple technical modifications to existing appliances and suggestions for modifying your habits of use where these may be wasteful. The measures I will discuss are cheap yet they involve substantial savings. They are always the logical first step to follow.

I will also suggest substitutes for appliances currently in your home. There are now many new energy-efficient appliances on the market that, despite their purchase costs, will pay for themselves in saved operating costs within a few years. As electric rates climb, the benefits of investing in new equipment accrue even more quickly and make such a move even more rational.

I will discuss the merits of using renewable energy devices when appropriate—devices that are powered by sunlight, firewood, or wind power instead of costly electricity. The natural energy flow of the wind or the sun, for example, can be used to perform several household heating and cooling tasks. Although several solar technologies are only marginally worth pursuing, several others are quite cost-effective.

It is important to become aware of the way you use electricity in your home. Cheap energy prices in the past and our national level of affluence have enabled most of us to pay little or no attention to our energy use. But at today's high rates, we need to acquire a sense of the relative costs of doing specific tasks electrically. Table 1-1 shows how long various electrical appliances will run on a dollar's worth of electricity. The lefthand column shows operating times based on an electric rate of 9¢ per kwh. The righthand column uses a rate of 12¢ per kwh, the anticipated average rate for 1988 to 1989.

The most obvious point to be gleaned from table 1-1 is that any form of electrical heating is very expensive. Running electricity through a high-resistance coil in order to heat air or water makes very little sense, and recognizing this is a first step toward trimming your monthly electric bill. The appropriate use of

Table 1-1

How Much Electrical Time Can You Buy for a Dollar?

Appliance	Running Time at 9¢ per kwh hr./min.	Running Time at 12¢ per kwh hr./min.
Baseboard heating system (1,800 sq. ft. house)	0:37	0:28
Central air conditioning (48,000 Btu, EER 8.4)	2:00	1:30
Clothes dryer (5,500 watts)	2:00	1:30
Water heater (5,500 watts—25 gal. per hr. recovery)	2:00	1:30
Water heater (3,800 watts—17.2 gal. per hr. recovery)	2:54	2:12
Electric oven (3,000 watts)	3:36	2:42
Range top burner (large, 1,700 watts)	6:30	4:54
Dishwasher (1,300-watt heat element)	8:24	6:18
Room air conditioner (10,000 Btu, EER 8)	8:48	6:36
Electric frying pan (1,200 watts)	9:12	6:54
Iron (1,200 watts)	9:12	6:54
Toaster (91,100 watts)	10:00	7:30
Coffee maker (automatic, 800 watts)	14:00	10:30
Clothes washer (500 watts)	18:00	13:30
Refrigerator (14 cu. ft., frost-free, 600 watts)	18:00	13:30
Freezer (15 cu. ft., frost-free, 450 watts)	25:00	19:00
Furnace blower (450 watts)	25:00	19:00
Blender (250 watts)	44:00	33:00
Dishwasher (no electric element, 250 watts)	44:00	33:00
Window fan (20-inch, 200 watts)	56:00	42:00
Stereo (solid-state, 120 watts)	92:00	69:00
Color TV (solid-state, 100 watts)	111:00	83:00
Light bulb (75 watts)	147:00	110:00
Sewing machine (75 watts)	147:00	110:00

electrical energy is that of running motors and solid-state equipment. Thus, for $1, you will probably get a year's use of your blender—44 hours—or 18 hours of work out of your automatic clothes washer. You will only get about 2 hours from your water heater, however, and only about 3½ hours from your oven for the same dollar. It can easily cost you $1.00 to $1.50 to cook a Sunday roast,

and it will cost $2.00 to $2.50 at future rates. Clearly, the sooner you begin to reduce your electrical consumption in these areas, the better.

There was a time when electric rates were actually dropping. This will probably never occur again. The promotion of all-electric homes by utilities in order to generate new demand and hence greater revenues is a thing of the past. In the future, new revenues can only come from higher-priced power, which will drive electricity use down further. The spiral of rising prices for electricity is here to stay, and despite arguments that other forms of energy—especially natural gas and fuel oil—will also rise in price, it is unlikely that the cost of either of these fuels will even come close to the cost of electricity in this century. In fact, the price of electricity will rise faster than the price of any other form of energy, making it even more expensive relative to other fuels than it is today. Based on these trends, it is almost mandatory that all but the most affluent families begin to control their use of electricity in a wise and rational way.

Table 1-2 shows how much heat energy is contained in one dollar's worth of various fuels, based on typical fuel prices of today. Thus, one dollar's worth of electricity contains about 38,000 Btu (a Btu—a British thermal unit—is roughly equivalent to the heat from a wooden match). Notice that the fuels that are most commonly used as alternatives to electricity for space and water heating (i.e., #2 heating oil and natural gas) provide significantly more heat per dollar than electricity does. Since the amount of heat you can actually *use* is determined by the efficiency of your heating equipment, and because oil- and natural gas-fired systems are less efficient than electrical systems, the proportion of *usable* heat energy you get per dollar with oil or gas versus electricity is actually less than the table indicates. Even when you take this into account, however, these fuels still deliver more heat per dollar than electricity does, and the use of newer, more efficient gas and oil appliances makes these fuels an even better bargain compared to electricity.

Table 1-2

Household Fuels: Energy Content per Dollar

Fuel	Unit Cost	Btu per Dollar
Electricity	9¢ per kwh	38,000
Liquid propane (LP)	$1.35/gal.	67,000
Kerosene	$1.45/gal.	93,000
#2 Heating oil	$1.20/gal.	116,000
Natural gas	$7.00/1,000 cu. ft.	146,000
Wood	$120.00/cord	167,000
Coal	$140.00/ton	193,000

A Few Things You Should Know

The following chapters give detailed suggestions for cutting your electric bills. Before getting to them, however, let's cover a few basic matters.

How to Read an Electric Meter

If you are committed to lowering your electric bill, you may want to take regular readings of your electric meter in order to monitor the impact of your energy-conservation efforts. Here's how to read the meter:

1. Read from left to right.

2. Always refer to the lower of two numbers when the pointer is between them. Thus, the meter in figure 1-2 shows 9,449 kilowatt-hours.

3. The reading is cumulative. In order to determine how much electricity you're using, take readings at two separate times and subtract the earlier figure from the later reading. For example, if your meter now reads 44,282 kwh whereas 30 days earlier it read 43,100 kwh, you have used 1,182 kwh during those 30 days.

ELECTRIC METER

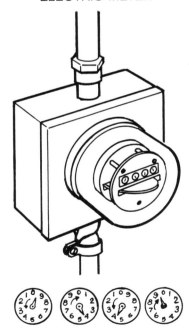

Figure 1-2. Reading your home's electric meter lets you keep track of your progress in cutting the amount of electricity you use. It also lets you check the accuracy of the electric bills you receive.

Energy Labels on Appliances

Air conditioners, water heaters, dishwashers, clothes washers, freezers, and refrigerator/freezers are now sold with energy labels that indicate the energy required to operate them. As the cost of electricity increases, the operating cost of an appliance becomes very important in your purchase decision. A new refrigerator/freezer that uses 1,100 kwh a year at 9¢ per kwh will cost $63 less to operate per year than a model of equivalent size that uses 1,800 kwh. If the efficient model costs $750 to buy, for example, and the less efficient model costs $550, the efficient model will earn back its additional purchase cost in 3 to 4 years. Over 15 years of operation—the average life expectancy of a refrigerator—it should save you at least $1,000 compared to the cheaper model. So clearly, you should compare energy labels and save.

Energy Jargon

Several terms are used repeatedly in the following chapters.

The *watt* is a measure of electrical energy; the *kilowatt-hour* (kwh) is the consumption of 1,000 watts sustained over a 1-hour period. Thus, a single 100-watt light bulb burned for 10 hours will use 1 kwh of electricity.

A *British thermal unit* (Btu) is a measure of heat energy. If 1 kilowatt is converted to heat, it will provide 3,412 Btu of heat energy. This is about as much energy as it takes to heat 7 gallons of water from 50°F (the temperature of incoming town or well water) to 120°F. One gallon of fuel oil provides 108,000 Btu of heat when it is burned in equipment that is 75 percent efficient.

R-value is a measure of the effectiveness of an insulating material. High R-values indicate a high resistance to heat flow; thus, materials having high R-values hold heat well. Ceilings, walls, floors, and hot water tanks that are well insulated have high R-values and have much slower heat losses.

Heat loss occurs when a surface is poorly insulated or when it contains openings that allow heat to escape. Clearly, there is little point in heating an area that is poorly insulated; the heat will be quickly lost.

Heat load is the amount of heat a house requires. The load can be handled by a single heating system—for example, a baseboard heating system might provide all the heat for a house. Or the load can be divided between two or more systems—for example, a house might have a baseboard heating system supplemented by a solar greenhouse or a wood-burning stove.

Payback period is the length of time it takes for your investment in conservation measures or in new energy-efficient appliances to be returned to you in savings. If an insulating jacket for a water heater costs $20 and reduces your monthly electric bill by $10, it has a two-month payback period (20 divided by 10 equals 2).

Off-Peak Rates

There are several hundred electric companies in the United States, each with its own set of rates and charges. Most customers will pay a general rate

per kilowatt-hour while customers with electric heat, electric water heaters, or all-electric homes (i.e., people who buy large quantities of electricity) often pay a somewhat lower rate. You should check your electric bill or call your electric company to determine what rate you pay. Make sure that you are paying the lowest rate allowable based on the amount of electricity you use.

You should also inquire about "off-peak" rates. Electric companies frequently charge higher rates during times when demand on their generating capacity is highest, and they offer lower rates when demand is low. They do this to discourage the use of electrical appliances during the "peak demand" periods. A utility must install sufficient generating capacity to meet the demand for power during the peak demand periods, plus a bit more (i.e., the utility must have some reserve capacity). This can mean investment in generating plants that have far more capacity than is needed most of the time.

The utility would much rather have a constant demand for its power and thus get optimal use of its equipment. But from the customer's perspective, demand naturally rises and falls. A hot summer day is the time to use your air conditioner, and early evening is the time to take showers, wash dishes, etc. This basic conflict of interests led to the creation of two types of electric rates: on-peak rates (the rates charged during periods of peak demand) and off-peak rates (the rates charged during periods of low demand). If you can take advantage of off-peak rates, you will lower your overall electric bill.

Rates may rise and fall according to the season of the year and the time of day. Because high summer demand due to household air conditioning use can force electric companies to buy a power plant or two more than they would need during other times of the year, summer rates are often increased to help keep down demand and cover the extra plant costs. Reducing your electrical usage during the summer—for example, by using the low-cost cooling techniques discussed in this book instead of air conditioning—is thus a wise move.

Some electric companies also have various rates for various times of the day. For example, late afternoon and evening hours are usually high-demand periods. If you can avoid using much electricity during these hours and concentrate most of your electric usage in the morning and night, you could benefit from off-peak rates offered during the morning and nighttime hours. Bear in mind, however, that you must sign up for the off-peak rates—the utility company will not automatically give them to you. Also, if you do sign up for off-peak rates, your rates during the on-peak hours will go up.

Many electric companies do not offer off-peak rates to their customers, and many that do may require that you be a big user of electric power, say 1,000 kwh or more per month. In addition, off-peak billing usually requires that you have a special meter installed that will cost you at least several dollars extra per month. Nonetheless, gearing appliance use to off-peak rates can save you quite a bit on electricity bills.

A residential customer of Pacific Gas and Electric Company in the San Francisco area, for example, can take advantage of off-peak rates if he or she uses 12,000 kwh or more a year. During the summer months, on-peak rates

are around 14¢ per kwh for the first 44 kwh used and 22¢ for each additional kwh. These rates apply to use during the hours from noon to 6:00 P.M. Off-peak rates, available from 6:00 P.M. to noon, are about 5.5¢ for the first 176 kwh used and about 8¢ per kwh thereafter. There are slight differences in the winter rates, but they follow a similar pattern with much higher rates charged during the on-peak period than the off-peak period.

A cost-conscious consumer could plan the use of many appliances to fit the lower-rate periods. Thus, timers could be attached to various appliances so that these machines would come on late at night. You might arrange for late-night water heating, dishwashing, clothes washing, and possibly even cooking. As a money-saving strategy it is an excellent approach. Check with your electric company to determine if off-peak rates are available to you.

How Much Can You Save?

Table 1-3 can be used as a quick reference guide for estimating how much money you can save by following the steps outlined in this book. To use the table, you need to know how much you now pay per kilowatt-hour of electricity and how many kilowatt-hours you use each month. Your most recent electric bill should show your rate per kilowatt-hour and the number of kilowatt-hours used. If this information is not given, call your electric company and ask them for these figures. Keep in mind that there may be a seasonal variation in your use of electricity and that your electric company may charge higher rates during the summer months. If you receive a bill only every other month, be sure and divide the kilowatt-hours used by two.

Table 1-3 is read as follows: Find your rate per kilowatt-hour to the nearest cent in the left-hand column. Then find your approximate monthly use—rounded to the nearest hundred kilowatt-hours—in the top column. The intersection of the two columns is your projected annual savings in dollars based on the assumption that by following the advice in this book you cut your electric use by 50 percent. A family paying 9¢ per kwh and presently using 800 kwh each month could save $432, for example.

If your present kilowatt-hour use is somewhere between the increments given—for example, if you use 450 kwh per month—adjust the estimated savings accordingly. Thus, a family paying 9¢ per kwh and presently using 750 kwh per month could save about $400 a year if they reduced their electrical use by 50 percent. If you will be less ambitious, for example if you only aim to reduce your electric use by 25 percent, you should make the appropriate adjustment. For example, if the family we just referred to decided to reduce their electric use by 25 percent instead of 50 percent, they should cut the $400 projected savings in half, to $200.

Table 1-3

Estimated Annual Savings from Reducing Electricity Use 50 Percent

Electric Rate per kwh (¢)	Kilowatt-Hours Currently Used per Month (nearest hundred)										
	200	300	400	500	600	700	800	900	1,000	1,100	1,200
2	$24	$36	$48	$60	$72	$84	$96	$108	$120	$132	$144
3	$36	$54	$72	$90	$108	$126	$144	$162	$180	$198	$216
4	$48	$72	$96	$120	$144	$168	$192	$216	$240	$264	$288
5	$60	$90	$120	$150	$180	$210	$240	$270	$300	$330	$360
6	$72	$108	$144	$180	$216	$252	$288	$324	$360	$396	$432
7	$84	$126	$168	$210	$252	$294	$336	$378	$420	$462	$504
8	$96	$144	$192	$240	$288	$336	$384	$432	$480	$528	$576
9	$108	$162	$216	$270	$324	$378	$432	$486	$540	$594	$648
10	$120	$180	$240	$300	$360	$420	$480	$540	$600	$660	$720
11	$132	$198	$264	$330	$396	$462	$528	$594	$660	$726	$792
12	$144	$216	$288	$360	$432	$504	$576	$648	$720	$792	$864
13	$156	$234	$312	$390	$468	$546	$624	$702	$780	$858	$936
14	$168	$252	$316	$420	$504	$588	$672	$756	$840	$924	$1,008
15	$180	$270	$360	$450	$540	$630	$720	$810	$900	$990	$1,080
16	$192	$288	$384	$480	$576	$672	$768	$864	$960	$1,056	$1,152

Chapter 2
Hot Water

According to the U.S. Department of Energy, a typical home uses 450 gallons of hot water per week. A conventional 52-gallon electric water heater requires 6,350 kwh per year to provide this much water at a temperature of 140°F. About 80 percent of these kilowatt-hours are used to heat the water directly; the remaining 20 percent are used to make up for standby losses (i.e., heat lost from the water in the tank to the surrounding environment). At 9¢ per kwh, a typical family would spend $48 per month for hot water.

Basic Conservation Measures

If you carry out a once-only, low-cost conservation program on your existing electric hot water system, you can save from 40 to 50 percent of your current expenditure for hot water. Here's how:

Step 1: Reduce the Temperature Setting

Locate the thermostat on your hot water tank, and if it is set higher than 130°F, set it back to 130°F. If your system has not been adjusted to a lower setting already, it is probably heating water to as high as 160°F.

Try the 130°F setting for a week; if it is satisfactory, try 120°F for the following week. If this is acceptable, keep the thermostat there or experiment with even lower settings. If your thermostat does not indicate actual degree settings, turn the dial down to the next lower level. For many families, a water temperature of as low as 110°F is satisfactory. Automatic dishwashing may be a problem at these lower temperatures, but as I point out later, this can be dealt with.

If your water heater has two thermostats, follow the same procedure, lowering the settings of them both. The upper thermostat should be kept at about a 10°F higher setting than the lower one. For example, if you set back the lower thermostat to 120°F, set back the upper one to 130°F.

ELECTRIC WATER HEATER

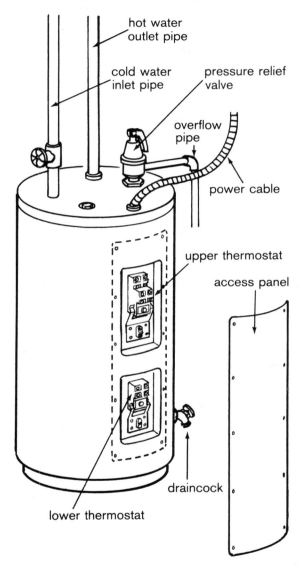

Figure 2-1. Water heater thermostats should be set to the lowest temperature you find acceptable for washing. Upper thermostats should be set 10°F higher than lower thermostats.

Step 2: Insulate Your Hot Water Tank

Unless you own a new energy-efficient hot water tank, your tank is probably poorly insulated and, thus, its heat losses are apt to be high. For $15 to $20, you can buy a 1½-inch-thick fiberglass tank jacket that is easily installed. It will reduce heat losses fairly well. I would recommend, however, that instead of using such a relatively thin tank jacket, you purchase a roll of 3½- or 5½-inch-thick foil-backed fiberglass and use this as a tank wrap. It is simple to install. Just wrap it around the tank sides and top and tape it together with duct tape. Make sure the thermostat(s) and the pressure relief valve are not blocked. Also make sure the foil side of your insulation faces out. The cost is about $15 to $20, but the insulating ability of 5½-inch-thick fiberglass is appreciably greater than that of a 1½-inch-thick jacket.

Step 3: Use Flow Restrictors

At hardware and discount stores, you can purchase a number of devices that restrict hot water flow at the tap, faucet, and shower head. Flow restrictors costing as little as 10¢ to 50¢ have been marketed during the past several years, but most manufacturers have elaborated on these small devices in order to make more money.

The less expensive washer-type flow restrictors may still be available at plumbing supply stores. They are easily installed at shower heads and faucets by merely unscrewing the outlet ends, inserting the restrictors, and screwing the original hardware back together again.

Aerator restrictors—which mix air with the water—can reduce the flow of water by about 60 percent. You will probably not be able to detect the reduced flow except on your monthly electric bill. Aerators sell for $2 to $5 each.

Special shower heads that reduce hot water output by 60 to 70 percent are also available for $8 to $20. An overall reduction of 30 percent in household hot water use is possible depending on the amount of reduction at each tap and the types of water-using appliances used within the household.

For more information about flow restrictors, see the section of part 2 titled "Energy-Efficient Water Heaters."

The above three-step conservation retrofit should require about 4 hours and an investment of $40 to $50. If you turn your water heater's thermostat from 140°F to 130°F, wrap 3½ inches of fiberglass insulation around the tank, and install flow restrictors that reduce hot water consumption by 27 percent—from 450 to 330 gallons per week, for example—you could reduce your electric hot water requirements from 6,350 kwh to less than 3,600 kwh per year. This is a reduction of 44 percent—a savings of over $250 per year. A turndown to 120°F, a 6-inch fiberglass wrap, and a flow restriction of 35 percent can provide an overall savings of 50 to 60 percent on water heating. Based on a $35 to

$50 outlay, you can expect a payback period of two to three months on your investment.

Other Conservation Measures

The three steps we have described will save you a great deal of money. You may not wish to go on any further in pursuit of hot water savings. If so, fine—proceed to the next chapter and feel confident you have made a huge dent in your monthly water-heating bill. On the other hand, if you wish to save even more, read on.

Step 4: Put a Timer or Manual Switch on Your Water Heater

Chances are, there are times each day when you don't use hot water. In fact, you may use it almost exclusively within a very restricted period. Why keep heating it for 24 hours? Let's assume there are four people in your house. Two of them shower every morning and two others shower or bathe in the evening. Dishes are done after dinner each day, and clothes washing is done two or three mornings each week. It would seem that there is probably no need to heat water 24 hours a day—your family uses most of its hot water during specific times of the day.

A timer or manual switch spliced into your hot water system's power line would allow you to restrict the hours that you actually heat water. This does not mean there will be no hot water during the periods when the heater is off. Unless you drain the tank just before switching it off, there will always be some water that is at least lukewarm in the tank. If you have insulated your tank well, the loss of heat from this water, once power is shut off, will be very slow—the water will remain quite warm.

Not only will a timer reduce the number of hours your water heater will draw electric power each day, it will offer the added advantage of allowing you to heat water during off-peak periods. If your local electric company offers lower off-peak rates—and if you qualify for them—timing your hot water systems to operate during the off-peak hours is an ideal cost-saving strategy.

For most families, a timer is recommended rather than a manual switch. No one will have to remember to turn the water heater on and off at given times each day; the process will occur automatically. Timers cost from $15 to $40. They vary according to the wattage of your system and the number of on/off cycles that can be set for each day. Be sure to match the timer's wattage rating with that of your water heater. Timers are available at discount, home improvement, and hardware stores, and at electrical and plumbing supply outlets.

Step 5: Insulate Your Pipes

If your hot water pipes are located in an unheated basement or crawl space, you should insulate them. If you have 30 feet of copper hot water pipe under your house, for example, you may lose heat equivalent to about 170 kwh through them each year.

Pipe insulation kits are available. Most kits are overpriced for what you get, however. I would recommend that you purchase 3½- by-15-inch fiberglass batts instead—the same product you can use to insulate your water tank. You may already have some extra fiberglass insulation around the house. Use it. Cut the fiberglass up the middle into strips 7½ inches wide. Wrap each strip around the pipe—foil facing out—and tape or staple the seam together. The cost for insulation and duct tape will be about 20¢ per linear foot, or $6 to $7 to cover a 30-foot length of pipe. This will provide you with two to three times the insulating value of even the best kits and will cost less. You can save between 50¢ to 70¢ per foot of pipe each year for a one-time investment of 20¢ per foot. A $7 investment could save an average family about $85 during the next five years. The payback period should be three or four months. This is a 1,200 percent return! Note that the hot water pipes are found by going to your water heater and locating the outlet pipe that feels hot to the touch.

Step 6: Fix Leaky Faucets

Between 175 and 250 gallons of hot water ($3 to $5 worth) can be lost through drips each month. Old washers are almost always at fault, and they are cheap and easy to replace. Turn off the incoming water valve under the sink or in the basement. Use a wrench and screwdriver to remove the faucet. Remove the old washer, get a replacement of the same size at your local hardware store, insert it, replace the hardware again, and turn the water back on.

Step 7: Use Less Hot Water

There are several techniques you can use to reduce hot water consumption. A shower almost always uses less water than a bath, and with flow restrictors, the difference is even greater. Dishes can be washed by hand using as little as 4 gallons of hot water. Automatic dishwashers typically use 12 gallons or more of hot water and draw additional power for the electric drying element. Always be sure to fill clothes washers to capacity. A full load means more efficient use of hot water. For both dishwashing and clothes washing, rinsing with cool water will save you money.

Shifting to More Energy-Efficient Equipment

Besides the conservation strategies we have listed, you should also consider trading up to energy-saving appliances or water heaters that use alternate fuels. Your options are given below:

Option 1: Convert to a Gas Hot Water System

The price of natural gas is rising and major increases are expected over the next few years. Nonetheless, the cost of electricity is so much greater than the cost of gas, and the anticipated price hikes for electricity are so large, that the shift from an electric system to a natural gas water heater makes economic sense.

Gas water heaters cost about $30 more than electric units, but operating costs are much less. Forty-gallon tanks are adequate for a family of four, if you use flow restrictors. A high-quality, energy-efficient gas unit with substantial foam insulation around the tank, an energy-saving thermostat, and an electronic pilot light sells for about $250 to $300. Efficient models are designed to transmit more heat energy into the water tank and to lose less heat to the surrounding environment during the heating process.

When you decide to purchase a new gas water-heating system, make sure you check its "transfer efficiency" rating, which indicates how successfully the unit transmits heat energy from the heating fuel into the water. The closer to 100 percent this figure is, the more hot water you will get for each dollar spent on fuel. A high transfer efficiency also means a faster recovery time (i.e., the time required to heat a tankful of cold water). This means a shorter wait between showers. Note that gas units generally have a quicker recovery time than equivalent electric water heaters.

If you now use an older low-efficiency electric water-heating system and convert to a high-efficiency gas system, expect to save up to 75 percent of your current cost of water heating. With flow restrictors and other conservation measures, expect to save even more. If you now pay $575 a year to heat water electrically, your annual bill should drop to below $200, even without conservation measures, and to $150 or less with conservation. The payback period for a $300 gas unit based on its annual savings to you of $300 to $375 is about 12 months. This means you will recover your entire investment in a year through operating cost savings. If you do not already have a gas line hookup at your house, there will be some additional costs associated with the conversion.

Note that, unfortunately, natural gas is not available in many areas of the

country. The cost of bottled propane is significantly higher than natural gas and makes conversion to a gas hot water system only marginally cost-effective.

For more information about gas water heating, see the section of part 2 titled "Energy-Efficient Water Heaters."

Option 2: Tankless Water Heating

Keeping a huge tank of hot water available for use at all times takes a great deal of energy. The Europeans, Japanese, and Australians are far more energy conscious than we Americans tend to be, and for years they have used "point-of-demand" or "point-of-use" devices to heat their water. The concept is a simple one. These water heaters do not store hot water in tanks for later use. Instead, they heat the water at the time it is needed.

Both electric and gas point-of-demand water heaters are now available in this country. Some are large enough to meet all the hot water needs of a typical family; others are smaller and are localized for use in a bathroom or kitchen. There are also units ideal for boosting water temperatures for dishwashing in homes where temperatures in tank water heaters have been set back to 110°F to 120°F. Point-of-demand water heaters (also known as tankless water heaters and instantaneous water heaters) all operate the same way. The heater comes on when you turn the hot water tap that is connected to the heater. Turn off the tap and the unit automatically shuts off again. Smaller units can be gas or electric; larger units tend to be gas fueled. In any case, gas is preferable because of its lower cost.

Conventional water heaters produce a given quantity of hot water over a certain time period. Most point-of-demand units vary in the quantity of hot water they produce per minute, depending on the temperature setting you choose. If you need very hot water, the volume or flow will be reduced; water of a lower temperature will arrive in greater quantities. Thus you can choose to get 3½ gallons of 95°F water or 1½ gallons of 150°F water per minute. Water of 115°F to 120°F is satisfactory for most uses—it should arrive at an average of 2 to 2½ gallons per minute. (Within limits, some of the newer tankless heaters allow you to escape the need to choose between quantity of water and temperature— see the section of part 2 titled "Tankless Gas Water Heaters.")

The main advantages of point-of-demand water heaters are that they never run out of hot water and the cost of operation is much lower than tank heaters— about 50 percent lower. A disadvantage is purchase cost. Tank heaters run from $150 to $400 while point-of-demand equipment usually ranges from $250 to $700, plus the cost of a gas hookup if you do not already have one. But because the durability and life expectancy of point-of-demand equipment is longer and the operating cost lower than conventional water-heating systems, they are an economically superior choice over your current electric system and are also worth considering over less efficient conventional gas systems. Point-of-demand equipment falls short only when compared to new high-efficiency gas-fired tank water heaters (option 1, above), which are cheaper to buy and install.

A conventional electric water heater may cost you about $575 a year to operate. Point-of-demand gas heaters could bring costs down to $200 or less a year. Point-of-demand water heaters normally are not used with flow restrictors, and since they have no tanks, insulation is irrelevant. Timers are also not applicable, and pipe lengths tend to be short and often located within the heated part of the house. Thus, there is little need for conservation improvements. An outlay of $800 for the equipment, installation, and gas line connection, if you don't already have it, would involve a payback period of about two years and an annual savings thereafter of 60 to 70 percent on your current cost of electric hot water.

Option 3: Purchase a Heat Pump

Heat pumps can provide an energy-conserving and cost-effective alternative to electric hot water systems. They are similar to refrigerators and air conditioners in operation. U.S. companies now make models that either connect up with existing hot water tanks or contain their own storage tanks.

Heat pumps extract heat from the surrounding air and transfer it to your water supply. How efficiently this is done depends on the temperature and humidity of the air. In 80°F air with 60 percent humidity, heat pumps attain high levels of operating efficiency. In colder situations, such as those prevailing from late fall through early spring, there is less heat to extract from the air and efficiency levels drop.

Heat pumps are often installed outside the house so that they can draw their heat from the outdoor air. In the northern sections of the United States, a heat pump must be installed inside the house to prevent the pump's condenser from freezing. Below an air temperature of 45°F, the pump shuts down and your conventional electric hot water system takes over. By placing the heat pump inside your heated house, this problem is avoided. However, since the pump draws heat from the space you are already paying to heat, there is a "cooling penalty" involved. This means that in the process of extracting heat from the air within the living space of your home, the heat pump actually increases the amount of heat you must provide to maintain a given comfort level. This is a relatively small problem if you use an inexpensive fuel such as firewood to provide the heat, but if you use electric heat, the cost can become a drain on your wallet.

Even under less-than-ideal conditions, the heat pump is a far better performer than electrical resistance hot water systems. In the latter, 1 kwh provides 3,412 Btu of heat energy to your water. By comparison, a heat pump inside your house can give you 8,000 to 9,000 Btu per kwh. As long as you are not

currently heating water with a fuel substantially cheaper than electricity, you will gain by using a heat pump to heat your water.

Assuming a relatively low efficiency from a basement-located heat pump where 6,824 Btu would be produced for each kilowatt-hour used to run the pump, a $288 savings (one-half) could be expected based on a $575 annual bill using a conventional electric system. This 50 percent average savings is consistent with numerous studies conducted around the country on heat pump performance. Since heat pumps cost between $700 and $1,200, the payback period would be two to four years at an electric rate of 9¢ per kwh.

High-efficiency conventional gas water heaters (option 1) and point-of-demand gas water heaters (option 2) can offer greater operating savings than heat pumps. If you do not have access to natural gas, however, heat pumps may be advisable: They should provide a savings of 50 percent or more on electrical water-heating costs. Conservation modifications can further reduce overall costs. Be sure to seek high-efficiency quality equipment if you decide to purchase a heat pump.

For more information about heat pumps, see the section of part 2 titled "Energy-Efficient Water Heaters."

Option 4: Preheat Your Water

One factor affecting the amount of energy needed to heat your water is the difference between the temperature of the cold water that enters your water heater and temperature of the hot water that the heater delivers. Reducing a water heater's thermostat setting reduces the consumption of electricity by diminishing this difference. Thus, well water or city water entering your home at 48°F needs to be raised 92°F in order to achieve a 140°F temperature level. A setback to 120°F saves 20°F and reduces your water-heating increment to 72°F.

It is also possible to reduce energy requirements by raising the temperature of the cold water before it is heated. Two possibilities exist here: (1) placing a "tempering tank" within the house in order to raise the water temperature to 60°F or more before it enters your heating equipment, and (2) using solar preheating.

A standard 80-gallon cold water holding tank is ideal for use as a tempering tank; it can be purchased at plumbing supply stores for about $100. The tank should be placed in an inconspicuous location such as in a laundry or empty closet. Since the temperature within your house normally ranges from 20°F to 40°F warmer than groundwater temperatures, the water in the tank is bound to gain some heat from its surrounding environment, thereby reducing the amount of fuel you'll have to use to heat the water.

Installation of a tempering tank is quite simple. The cold water line is cut before the incoming water reaches your water heater. A new section of pipe is added, running from the cut to the lower portion of the tempering tank. Another section of pipe is connected to the top of the tempering tank and run to the water heater. Thus, the tempering tank is, in effect, spliced into the cold water line: Incoming water must flow through the tempering tank before reaching the water heater.

Note that during the heating season, a tempering tank will increase your space-heating load slightly, since it draws heat from the air in the house. But the increased cost for space heating will be more than offset by the reduced water-heating costs. A tempering tank could save you $50 to $75 a year, depending on where you live and what type of water heater you own.

A variation on the tempering tank is the solar preheater. This device works much like a tempering tank, "preheating" water before it reaches the water heater. The only difference is that a solar preheater draws its heat from the sun rather than from warm air.

To set up a solar preheater, a water tank (30 to 42 gallons) is spliced into the incoming water line, so that water must pass through the tank on its way to your water heater. The tank is painted black and placed in a location where sunlight can strike it directly. This usually means putting the tank in a south-facing location behind glass. A solar greenhouse can be ideal for this purpose in all but the northernmost regions of the United States.

It is also possible to build or purchase an insulated box with a double layer of glass or Plexiglas glazing exposed to the south at about a 45-degree angle. Such a unit is known as a "passive-solar breadbox" or "batch water heater." It can be mounted on your roof or placed in your yard against the south side of your house, provided the sun is not blocked by trees or other buildings. Manufactured units are available with capacities of 20 to 55 gallons. Homemade units are, of course, more economical.

Batch heaters take you well beyond the savings of a tempering tank. In the North, a batch heater would probably have to be drained for the two to three coldest months to prevent freezing, unless it was placed within your house. The solar warmth that is lost during this "down time" would only be about 15 percent of the total amount of warmth the heater could receive annually from the sun. The remaining 85 percent available over the nine warmest months would still provide you with a substantial savings on water-heating costs. Studies indicate that up to 50 percent of annual hot water costs can be saved with a solar preheater. A savings of 30 to 35 percent is probably more realistic in colder areas.

The cost of constructing a batch heater should run about $350 to $700. Your actual savings will, of course, vary according to whether or not you carry out a conservation retrofit of your existing water heater. Assuming you conduct a complete conservation program on your electric hot water system and thus reduce annual costs by half, 30 to 50 percent of the remaining cost can be eliminated through the use of solar preheating.

For more information about passive-solar systems, see the section of part 2 titled "Residential Solar Water Heaters."

SOLAR BATCH HEATER

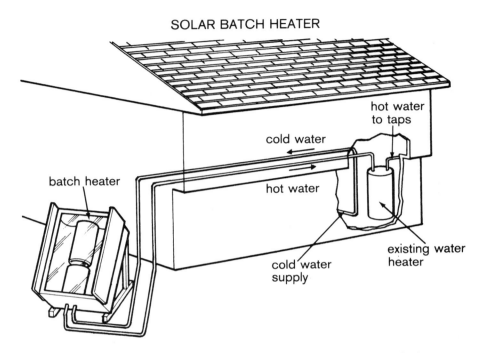

Figure 2-2. A "batch" solar water heater might trim your expenditures for hot water by 30 to 50 percent. The heater can be mounted on your roof or in your yard.

Option 5: Install an Active-Solar Water System

Solar preheaters are "passive-solar" devices: very simple, with virtually no mechanical components. The batch heater is the most cost-effective of these solar water systems, but unless you live in the South or Southwest, it will probably only meet about 35 percent of your annual hot water needs. An alternative is to use an "active-solar" system, which is more expensive and mechanically complex, but which may contribute 50 to 90 percent of your hot water needs. Active systems utilize control and monitoring devices, pumps, and other sophisticated hardware. Initial costs for these systems run about three to four times higher than for a heat pump ($2,000 to $3,500) and the maintenance costs are also apt to be greater. A system delivering the equivalent of about 2,500 kwh in the northern latitudes (more in the southern latitudes) would cost about $3,000.

If you now use 450 gallons of hot water per week and own a poorly insulated hot water tank, you are consuming about 122 kwh per week to heat water. The portion of this energy use that can be met with the sun's energy will vary according to your latitude, local climatic conditions, the orientation and tilt of your collectors, and how many collectors you have (i.e., the combined total area

ACTIVE-SOLAR WATER HEATING

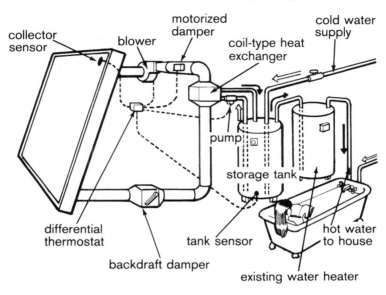

Figure 2-3. Active-solar water heating systems are mechanically complex, but one might provide up to 90 percent of the hot water your family needs.

of collector surface). If you live in the Northeast, for example, a standard solar water system should provide the equivalent of about 50 kwh of heat energy to your water each week. If, however, your water tank is well insulated and flow restrictors are used, hot water requirements are reduced and 50 percent or more of total demand can be met by the sun. If it cost you $575 per year to heat water before going to solar and installing flow restrictors, and if your water-heating costs were $300 afterward, your annual savings would be $275. The payback period on your $3,000 investment would thus be about ten years. If your electric rates are above 9¢ per kwh or go up during the next five years or if you live in one of the sunnier regions of the nation, you will get back your investment in less time.

Since the sun does not always shine—and even when it does, it may not provide sufficient heat to raise your water temperature to the desired temperature—a backup system of electrical resistance heat is often provided inside the water tank. This is the catch. You must be very careful that you do not purchase a system with either undersized collectors or an oversized tank. If you do, you will be back to consuming large quantities of electricity again: The electric resistance coils inside the tank will have to come on frequently to supplement the efforts of your incorrectly sized solar water system.

For more information about active-solar systems, see the section of part 2 titled "Residential Solar Water Heaters."

Recommendations

I will wrap up this chapter by summarizing my specific recommendations for water heating, given in the order of preference.

1. Convert to an energy-efficient gas hot water system and carry out a conservation retrofit. Expect to save 60 to 75 percent on your current outlay for hot water. Payback period: approximately 12 months.

or

2. Install a gas point-of-demand system. You should save about 60 to 65 percent of current outlay for electric water heating. Payback period: about three years.

or

3. Do a conservation retrofit on your existing electric hot water system. You will save 40 to 50 percent. Payback period: approximately one to two months.

or

4. Install a batch solar preheater and conduct a conservation retrofit on your existing water system. Save about 65 to 70 percent. If you convert to an energy-efficient gas tank (or tankless) system as well, expect to save 80 to 90 percent on hot water costs. Payback period: one to four years.

or

5. Install a heat pump and do a conservation retrofit. Savings of 65 to 75 percent. This is a more desirable option in warm southern regions. Payback period: about two to four years.

or

6. Purchase an active-solar hot water system. Adopt conservation measures where possible—especially flow restrictors. Expect to save 50 to 60 percent. Payback period: three to ten years, depending on where you live and what equipment you select.

Chapter 3
Lights

A typical American family consumes 1,200 kwh or $108 worth of electricity each year to provide interior lighting. Because lights are widely dispersed and because bulbs vary in wattage, type, purpose, and life expectancy, reducing electrical use for lights seems a difficult task. Don't be misled, however. It is, in fact, relatively easy to reduce energy use for lighting, and a 40 percent savings can be made with minimal effort on your part. Understanding a few basic facts about lighting is an important first step to cost reduction.

The wattage of a light bulb is a measure of how much electrical power it will consume—not how much light it will give you. The brightness of a bulb or how much actual light you get is measured in lumens. The more lumens you get per watt, the more efficient your bulb. Both lumens and watts are given on the bulb package. If you do a little basic calculation (divide lumens by watts) when you buy bulbs, you will realize that efficiencies vary substantially. The more lumens per watt, the more light you will get for each watt of electricity you buy.

Higher-wattage bulbs tend to be more efficient (more lumens per watt) than low-wattage bulbs. A 75-watt bulb actually gives you 68 percent more light than three 25-watt bulbs, even though both use the same amount of power. A single 100-watt bulb gives off 20 percent more light than two 60-watt bulbs, and it uses less power. If you have fixtures with two or more low-wattage bulbs, converting to fewer, higher-wattage bulbs can give you the same amount of light while using less electricity. (Be sure not to exceed the acceptable wattage for a particular fixture. Some lamps, for example, are designed for 60-watt bulbs or smaller: putting higher-wattage bulbs in such lamps can be dangerous.)

The greater efficiency of higher-wattage bulbs doesn't mean you should go around the house putting these bulbs in all the fixtures that are designed to accept them. On the contrary, people often overlight an area; conversion to lower-wattage bulbs is advisable. If you currently have a 100-watt bulb in a hallway, for example, you might experiment with a less powerful bulb. A 75-watt bulb, even a 60-watt or 40-watt bulb, may provide all the light you need there. You will trade a slight decrease in illumination for a reduction in your electric bills. You should use this strategy throughout the house—it can lead to significant savings.

The life expectancy of a bulb should also be considered. Long-life bulbs are less popular now than several years ago, but they are still being sold. Often

people assume that buying a long-life bulb is a better deal. The truth is that long life is achieved by reducing the bulbs' efficiency (i.e., the number of lumens per watt). A 765-lumen, 60-watt long-life bulb should last 2,000 hours and, at 9¢ per kwh, it should use $10 worth of electricity over its lifetime. It will produce 1,417 lumens for each cent of operating cost. Two 860-lumen, 60-watt bulbs, each with 1,000 hours of life, will give 1,593 lumens or 12.4 percent more light for the same operating cost. Thus, using the short-life bulbs represents a potential savings of 12.4 percent. When we take the purchase price of the bulbs into account, this figure is reduced to 6 percent—small, but every little bit helps. As the saying goes, a penny saved is a penny earned.

It is informative to compare the respective costs of the incandescent bulb options available today. Table 3-1 shows the cost of receiving one million lumens from each of five bulb types. Notice that there is very little difference in the cost of providing illumination with long-life, standard white, and so-called miser bulbs. The latter are designed to run on fewer watts but their higher purchase price and short life span actually make them more costly than standard white bulbs. Three-way bulbs allow you to choose different levels of illumination, but they do so at the expense of efficiency, and because their purchase price is relatively high, they are an unwise choice for most purposes.

Table 3-1

Cost per Million Lumens of 100-Watt Rated Bulb Types

Bulb Type	Rating	Life Expectancy (hr.)	Purchase Price ($)	Cost per Million Lumens* ($)
Three-way	100 watts† 1,640 lumens	1,200	2.19	6.60
Long-life	100 watts 1,585 lumens	1,500	0.75	5.99
"Miser"	95 watts 1,710 lumens	750	1.10	5.87
Standard white	100 watts 1,710 lumens	750	0.75	5.85
Reflector	50 watts 1,710 lumens	2,000	5.85	4.35

*Includes cost of bulbs based on their average life expectancy plus operating cost at 9¢ per kwh.
†50-100-150-watt bulb operated at 100-watt rating only.

Reflector bulbs are a less expensive way to provide household lighting. They are appropriate for illuminating a specific work area such as a kitchen counter, a sink, or a desk. They are 50-watt bulbs that cast light roughly equivalent to most 100-watt bulbs. Conventional bulbs transmit light energy in all directions, which is fine if your purpose is to light a large area. By comparison, the neck region of reflector bulbs is coated with a reflective surface so that virtually all of the light is beamed in one direction, thereby producing strong illumination for a relatively small area. Even with their high purchase cost, reflector bulbs are an ideal way to reduce electricity use. They run about 25 to 35 percent less in total cost per given amount of illumination than the other bulb types mentioned above.

The most economical choice of all is the circular fluorescent bulb that screws into a conventional incandescent light fixture or lamp. Table 3-2 shows the cost of using this type of bulb compared to other options. The combined purchase price and operating cost of the three incandescent options is approximately the same, but the screw-in fluorescent bulb provides light for about half this cost despite the extremely high purchase price of the bulb. The savings result from both the low wattage drawn by the bulb and from the long life expectancy of the bulb. Since the bulb can be used in regular lamps and fixtures, an additional savings comes from not needing to invest in special fluorescent fixtures, which are quite expensive. Fluorescent lights are far more efficient and long-lasting

---

Table 3-2

Cost per Million Lumens: Fluorescent versus Incandescent Bulbs

Bulb Type	Rating	Life Expectancy (hr.)	Purchase Price ($)	Cost per Million Lumens* ($)
Long-life	60 watts 820 lumens	1,500	0.75	7.20
Standard white	60 watts 855 lumens	1,000	0.75	7.19
"Miser"	60 (55) watts 855 lumens	1,000	1.10	7.08
Circular fluorescent	22 watts 870 lumens	12,000	13.99	3.60

*Includes cost of bulbs based on their average life expectancy plus operating cost at 9¢ per kwh.

than standard incandescent bulbs. A 40-watt fluorescent bulb is twice as efficient as a 100-watt incandescent bulb, it gives more light, and it lasts up to 20 times longer.

If economics alone were the grounds for choosing, fluorescent bulbs would clearly be the best option. However, most such bulbs provide an inferior form of cold light that is psychologically depressing. The flickering and the humming they produce cut their desirability so much that I am not willing to recommend them. Some studies even indicate that cell damage and cancer may be associated with their use. There are fluorescent lights now available that can provide a more balanced "full spectrum" light. This means they are warmer and less harsh on the eyes. They are designed to approximate natural light and help the body produce vitamin D in order to resist respiratory diseases. In the Soviet Union, these lights are used in schools to help children avoid colds. I personally do not use fluorescent lights, but if you feel comfortable with them, they can save you a great deal of money on your electricity bill.

Low-voltage incandescent bulbs are another relatively new option. These bulbs contain transformers or are connected to voltage-transforming devices. The transformers reduce 120-volt house current so that it enters the bulb at 5 or 12 volts. The bulbs themselves have small filaments and generate potent and controlled beams of light, suitable for illuminating work areas. However, in focusing light so well, the bulbs create a problem: they are somewhat hard on the eyes. Based on this drawback, low-voltage bulbs are probably less desirable than reflector bulbs.

Conservation Measures

Now that you know a bit more about light bulbs, here are some ways to save electricity when using them:

Step 1: Use Daylight whenever Possible

Nothing beats natural light for eye comfort. For daytime tasks such as reading, writing, studying, or sewing, a window can provide the ideal illumination. If you use electric lights during the day while keeping curtains or draperies partially drawn, consider how you might relocate various activities and tasks to areas where daylight is available and how the amount of natural light entering the house can be increased.

For more information about windows, see the section of part 2 titled "Windows, Skylights, and Related Devices."

Step 2: Turn Off the Lights

My wife grew up on a remote cattle station in Australia. Electric lights were operated from batteries that were charged periodically by a diesel generator. As the evening wore on, the batteries lost their charge and the lights gradually dimmed. A bulb needlessly left burning ultimately meant less light for someone to read by. If we could experience a week or two in similar circumstances, our electric light consciousness would be transformed forever.

As electric rates climb, we may become more sensitive to the cost of leaving lights on needlessly. If an area is to be unoccupied for more than 15 minutes, learn to turn off the lights. (Shutting off lights for shorter periods is not worthwhile since the energy savings will be offset by a reduction in the life of the bulbs. Turning a bulb on and off frequently is hard on the bulb's filament.)

Step 3: Install Light Dimmers

Household areas often lend themselves to different lighting levels at different times. An end-table lamp used for reading may require a 75-watt bulb, but 25 watts may be adequate when watching TV. A 60-watt bulb may be required over a kitchen counter during meal preparation, but 25 watts may be fine for providing minimal late evening illumination. Solid-state dimmers allow you to "turn down" your lights to suit the task and mood of the moment. These devices are inexpensive and also extend bulb life. (Do not use them with fluorescent lights, however—this could cause a fire.)

Step 4: Match Bulbs to Use

Ask yourself what task each lamp or light fixture in your house serves and consider the optimal bulb to meet that need. A desk light of 100 watts is probably too bright. Replace it with a 75- or 60-watt bulb or, better yet, with a 50-watt reflector bulb. Keep in mind that light-colored surfaces reflect more light, and shades that properly direct and reflect light allow you to get the most illumination from your bulb.

Overhead and ceiling lights are meant to provide general illumination rather than focused lighting for close work. You may find that, for most purposes, you really don't need high-wattage general area lighting. Instead, use a lamp to concentrate light on your work area and turn off overhead lighting.

Don't overlight. A 4- or 7-watt night-light will reduce electrical consumption.

So will a timer for security lighting: Instead of burning all day and night, your lamps will come on only during predetermined hours.

For more information about timers, see the section of part 2 titled "Timing Devices."

Step 5: Substitute More Efficient Light Bulbs

A single higher-wattage bulb is cheaper to use and more efficient than several lower-wattage bulbs. A frosted bulb provides more light than a soft-white bulb of equivalent wattage. A 50-watt reflector bulb can concentrate as much illumination as a 100-watt standard bulb, yet it costs only half as much to operate. Use short-life, frosted, and reflecting light bulbs of the appropriate wattages—you'll save money.

Step 6: Clean Light Fixtures and Dust the Bulbs

Dirt absorbs light. A periodic cleaning around lights and fixtures means more usable lumens will get to where you need them.

Note: "Light buttons" plug into your existing lamp sockets. They are supposed to reduce power costs by 50 percent and extend bulb life. These little devices sell for $1.50 to $4.00 and present a fire hazard. They extend bulb life by reducing the amount of light produced by the bulb and actually have a negative impact on bulb efficiency. This means you'll pay more for less light. Don't buy them. Don't use them.

Another Option

Most daylight will enter your house through normal vertical windows, of course. But you may also want to consider installing a "window" or two in your roof. Daylighting via skylights offers some potential energy savings. Properly designed skylights that are oriented to capture some solar heat energy during winter months and that also act as summer heat vents can offer a cost-effective, psychologically pleasing source of direct and diffuse light to your home's interior. Their major disadvantage is winter heat loss during the night and on cloudy days. If they are fitted with convenient-to-operate insulating shutters, however, heat loss will be reduced, in which case skylights can be an effective supplement to electric lighting.

For more information about skylights, see the section of part 2 titled "Windows, Skylights, and Related Devices."

Recommendations

Table 3-3 shows the electric lighting used by a typical American family. Table 3-4 reflects my specific recommendations for this family: It shows the types of bulbs that can be used in each room and the number of hours they might be used daily. By following these recommendations, the family could reduce its power consumption for lighting from 3,475 watts to 2,146 watts.

The savings shown in table 3-4 equal 1,329 watts per day. This constitutes a 38 percent reduction. An annual lighting bill of $108 would be reduced by $41. In no instance would the family have to sacrifice comfort, and in some cases they would actually improve the efficiency and the level of illumination.

Turning lights off costs nothing—the payback is instant. A dimmer costs $5 to $8. The substitution of natural daylight is also free, except if you elect to install new windows or skylights.

A slightly more rigorous pursuit of conservation and daylighting could increase the overall savings on the family's annual lighting even further. Over the coming decade, these no-cost and low-cost changes should make no noticeable difference to interior comfort and illumination, but they could save the family $400 to $500.

Table 3-3

Electrical Light Use by Typical Family before Conservation

Area	Bulbs (number and wattage)	Daily Use per Bulb (hr.)	Total Watts
Bathroom	1 60-watt ceiling bulb	2	120
Bedrooms (2)	4 60-watt lamp bulbs	2	480
Dining room	1 60-watt ceiling bulb	3	180
	1 60-watt lamp bulb	3	180
Hallway	1 40-watt night-light	6	240
Kitchen	1 100-watt bulb over sink	4	400
Laundry	2 60-watt ceiling bulbs	2	240
Living room	2 60-watt ceiling bulbs	6	720
	1 60-watt lamp bulb	4	240
	1 75-watt lamp bulb	5	375
Study	1 100-watt lamp bulb	3	300
Total			3,475

			Table 3-4

Electrical Light Use by
Typical Family after Conservation

Area	Bulbs (number and wattage)	Daily Use per Bulb (hr.)	Total Watts
Bathroom	1 22-watt ceiling bulb	2	44
Bedrooms (2)	4 60-watt lamps (and natural light)	1.5	360
Dining room	1 60-watt ceiling bulb (with light turned off after meals)	1.5	90
	1 60-watt lamp	1.5	90
Hallway	1 7-watt night-light	6	42
Kitchen	1 50-watt reflector bulb over sink	4	200
Laundry	1 100-watt ceiling bulb	2	200
Living room	1 100-watt ceiling bulb	5	500
	1 60-watt lamp (with dimmer)	2 at 25 watts	50
		2 at 60 watts	120
	1 60-watt lamp	5	300
Study	1 50-watt lamp (reflector bulb)	3	150
Total			2,146

Chapter 4
Refrigeration

Refrigerators may hold your bread and butter, but they *are* the electric companies' bread and butter. Most users of electricity do not have electric heat, and many do not have electric hot water systems or electric ranges. Virtually everyone owns a refrigerator, however. A spokesman for a New York electric company recently noted that a 50 percent rise in electric rates would have only a limited impact on electric power consumption since the biggest power user in most homes is the frost-free refrigerator, and there is very little that people can do about it.

The spokesman may have exaggerated slightly. Cutbacks in power use for refrigeration are unquestionably more difficult than in other areas, but some savings are possible nevertheless, especially if you are strongly committed to reducing your monthly electric bill. Refrigerators are not inherently wasteful devices, but our habits and lifestyles have made them less efficient over time.

The refrigerator, like the heat pump described earlier, operates by extracting heat from the air at one location and transferring it elsewhere. In this case, heat is drawn from the air inside a sealed and insulated box, and it is sent to the surrounding environment. The cost of doing this varies according to the efficiency of the refrigerator's motor, how good a job the refrigerator does keeping cold air in and hot air out, and how quickly hot air is dissipated from the area directly around the refrigerator. Frost-free models have the disadvantage of actually producing heat inside the box to abolish the frost, and thus their motors need to work harder to keep things cool inside.

Cutting Refrigeration Costs

Here are some steps you can take to reduce the cost of refrigeration:

Step 1: Do a Conservation Retrofit on Your Refrigerator

Get an outdoor thermometer and place it inside your refrigerator. If the temperature reading is lower than 40°F, reset the refrigerator's temperature control to the next highest (warmer) level. If the temperature is still below 40°F after a few hours, go up another notch. Like hot water heaters that are set too high and needlessly waste energy, refrigerators frequently are set too low. A higher setting will save you money.

Unplug your refrigerator. Pull it away from the wall. If there is a layer of dust at the back and around the motor area, clean it away. A narrow brush or vacuum cleaner will do the job. Clean away any dust around the grill area under your refrigerator, also. When you shift the refrigerator back into place, leave 4 to 6 inches between it and the wall. This and the removal of dust will allow waste heat from the refrigerator's motor and coils to dissipate more readily. Better circulation means your motor works less.

Check your door gaskets. Close the refrigerator door on a dollar bill. (Do this with the freezer, too, if it has a separate door.) The more difficult it is to remove the dollar, the better. If the dollar comes out easily, so will the cold air inside the refrigerator. Repeat the procedure at four or five locations around the door—the seal may be tight in one place but loose in another. If the bill offers

little resistance, you probably need to replace the gasket. Also check for cracks, breaks, or brittleness in the gasket. If any of these problems exist, replace the gasket.

Gaskets are available in hardware stores. Use a putty knife to remove the old gasket, then cement the new gasket in place. Gaskets for some refrigerator models can be tricky to install. Check with your local serviceperson about your particular brand. After installing a new gasket, check it with a dollar bill.

Sometimes the refrigerator door itself may be slightly out of alignment, causing gaps between the seal and the refrigerator. Years of use can cause a door to sag. Often a door will become loose at one of the hinges. Check this by lifting the door to see if there is any play. If so, simply tightening the hinge screws should remedy the problem.

REFRIGERATOR GASKET

gasket

Figure 4-1. If a refrigerator gasket has deteriorated, it should be replaced. New gaskets are available in hardware stores.

Step 2: Change Three Habits of Refrigerator Use

The purpose of your refrigerator is to keep the contents cold. If you put hot items inside, the refrigerator will take more energy to cool them and the rest of its contents. Let hot items get cool outside your refrigerator before putting them away.

Keep the contents of your refrigerator to a minimum. If the shelves of your refrigerator are commonly filled with empty bottles, leftovers from weeks ago that will never get eaten, or items that do not really require refrigeration, you are wasting energy and money. If you drink one bottle of beer every day or two, there is no need to keep twelve bottles cold. Look things over more often; remove things that do not really need to be there.

Do not stare into a wide-open refrigerator for interminable periods of time. Learn to make quick decisions or think about what you might be looking for before you open the door. It takes a lot longer for cold air to build up inside a refrigerator than it does for the cold air to escape when you swing open the door.

Steps 1 and 2 above could reduce your overall refrigeration bill by up to 30 percent, depending on prior conditions and habits. An average household should save between $25 and $42 per year. The older and less efficient your equipment, the more you will gain by carrying out conservation. Table 4-1 gives you an idea of what it currently costs to run your refrigerator/freezer according to its size. In the future as rates increase, the cost of operating your refrigerator could easily climb to $150 to $200 a year.

Table 4-1

Home Refrigerator/Freezer Operating Costs*

Model Size (cu. ft.)	Average Operating Cost	
	Month ($)	Year ($)
16.5–18.4	10.08	121.00
14.5–16.4	9.50	114.00
12.5–14.4	9.00	108.00
10.5–12.4	6.83	82.00
11 (manual defrost)	2.92	35.00

*at 9¢ per kwh.

Step 3: Buy a More Efficient Unit

If you own an inefficient refrigerator, you may want to consider retiring your old model and purchasing a new one. In 1950, the average 14- to 16-cubic-foot refrigerator used only 750 kwh a year. Then we began to add more and more energy-using features, and by 1975 the figure had climbed to 1,800 kwh. If you own a 1975 or earlier model, you may be paying $150 a year to keep it going. By 1980, energy-efficient frost-free 14- to 16-cubic-foot refrigerator/freez-

ers used only 900 kwh per year. There are now several energy-efficient refrigerators being sold in the United States. Some will operate for as little as $70 to $80 per year.

A savings of 50 percent or more is possible by converting to newer equipment. Look for a refrigerator having at least 2 to 2½ inches of foam insulation in its walls, tight gaskets, efficient compressors, and "power-saving" switches that can cut off the antifrost elements inside when the humidity is low (frost builds up fastest in humid weather). These switches alone can save you roughly 16 percent. You may also wish to look at manual-defrost refrigerators. They cost about half as much as the frost-free models to operate, although they require more work on your part.

The payback period for a new refrigerator can be as short as six years at 9¢ per kwh and four to five years at anticipated higher rates of 12¢ per kwh. This payback period is based on an expected savings of 50 percent on current costs.

Note that federal law requires that new refrigerator/freezers and other major appliances excluding ranges must carry labels showing their estimated annual operating costs. The estimated costs shown on these "Energy Guide" labels are based on U.S. government standardized tests. Bear in mind that the way you use an appliance will influence how much you actually pay each year.

Energy labels also help guide you by listing the lowest and highest estimated annual operating costs for similar models of the same type of appliance. If you are looking at a medium-size refrigerator, for example, its label will show its estimated operating costs as well as the costs for the most-efficient and least-efficient refrigerators of the same size. Energy labels are a useful consumer tool; study them carefully when purchasing appliances.

For more information about energy-efficient refrigerators, see the section of part 2 titled "Refrigerators and Freezers."

Freezers

It is very expensive to keep a separate freezer, and in many cases it is not worth doing. Major savings on the purchase of lower-cost or bulk food items or the preserving of garden produce, may make owning a freezer worthwhile, but not necessarily.

If you feel a freezer is justified in your circumstances, be sure gaskets are tight. Put the temperature setting at no lower than 0°F. Open the door as infrequently as possible. Place your freezer in a cool location and away from direct sunlight. Keep it full (this helps store up cold in its mass) and make sure air circulation is adequate by leaving some space at the back and sides, if possible.

Energy-efficient freezers do exist. Models ranging from 13.5 to 15.4 cubic feet can cost as little as $45 a year to operate at 9¢ per kwh. Always opt for a chest freezer instead of an upright. Uprights dump substantial amounts of cold air every time they are opened—the cold air simply pours out of the open door. Since cold air tends to sink (it is heavier than hot air), this is much less of a problem in chest freezers—the cold air tends to remain inside the freezer when the door is opened.

If you already have a freezer and wish to keep it going, wrap additional insulation around the sides. Styrofoam board 2 inches thick taped to the sides of the freezer will reduce losses. Consider the possibility of an annual "down time"—a period during the warmer months when your freezer is emptied and turned off. A freezer's big energy consumption time is during summer months. An annual summer fallow period will do wonders for the yearly overall operating cost of your freezer.

For more information about freezers, see the section of part 2 titled "Refrigerators and Freezers."

Recommendation

If you wish to keep your existing refrigerator, carry out the conservation steps described above. If the model you now use is more than eight years old, you may want to look into a new energy-efficient model. By cutting your annual costs for refrigeration 50 percent, you could achieve a payback period of just six years at 9¢ per kwh.

Chapter 5
Cooking

Electric ranges are big energy users. An average family can expect to use about 24 kwh per week to prepare food. At today's electricity costs, an average family pays about $9.50 a month to operate its range.

Cutting Cooking Costs

Cooking with an electric range is wasteful. The ultimate solution is to get rid of it. The truth is that there are few modifications you can make to your

existing stove to boost its efficiency. However, if you are determined to keep your electric range, you can still reduce your electric bills by changing your cooking habits. Observe some of the following tips:

1 Match pots and pans to the size of burners. A pot should completely cover the burner without extending more than an inch beyond. Thus, maximum heat is conducted to the pot and little heat is wasted. A small pot on a large burner is especially wasteful. Use smaller burners whenever possible—they use 50 to 60 percent of the electricity that larger burners do.

2 When boiling water, use only as much as you need. If you want to make three cups of tea, pour three cups of water into the kettle. Once you acquire the habit of doing this, it will seem natural and not as inconvenient as it may sound.

3 Plan meals so that several dishes, an entire meal, or even several meals can be placed in the oven at one time. You can refrigerate or freeze the extra and reheat small portions later.

4 Thaw frozen foods to room temperature. This will save about one-third of the energy needed to cook them.

5 Keep stove-top reflectors and the bottoms of pots clean for more efficient heat transfer.

6 Cover pots. This will save 20 percent of the energy required to boil water or cook food.

7 Stagger shelves and dishes in the oven to optimize heat circulation.

8 Preheat selectively. If the food you are cooking is to be in the oven for over 45 minutes, preheating is probably unnecessary.

9 Turn off burners and the oven a bit early and let the food cook on stored heat. An oven uses 27¢ worth of electricity in 3 hours. If you turn your oven off 15 minutes early, you could save a few cents on every meal.

10 Do not peek under lids or open the oven door needlessly. You can save up to 20 percent on energy per meal.

11 Do not use your oven's self-cleaner. It uses great quantities of electricity. If you must use it, do so immediately after you have used the oven to take advantage of the heat already there.

12 Whenever practical, cook on the top of your range rather than inside your oven. The burners are in direct contact with the pot or pan and the food inside it, whereas an item placed inside the oven is heated indirectly: the oven heats the air, then the air heats the food. This is less efficient.

13 Broil rather than bake or roast. You will use heat to directly cook your food rather than to heat space around your food.

14 Use a clock or timer to tell you when the food is supposed to be done. This saves excess cooking time and also avoids heat losses due to opening the door to check on the food.

15 Replace torn, broken, or cracked seals on your oven door.

Conversion Options

The collective impact of following the tips I've just described can be a 35 percent reduction in annual cooking costs. Even greater savings are possible through the following options:

Option 1: Convert to Energy-Saving Supplemental Appliances

The *pressure cooker*—essentially a sealed metal pot used on a range-top burner—offers a fast, cheap means to cook much of your food. Using only a small amount of water—½ cup or so—in a sealed cooker, it steams your food under 15 pounds of pressure. This high pressure in turn drives up the boiling point of water to 250°F, allowing for much higher than normal cooking temperatures and an extraordinarily fast cooking time. Most vegetables can be cooked in 1 to 7 minutes, for example. In most cases, the quality of the food is much higher and the retention of vitamins much greater with pressure cooking. When a pressure cooker is used on a daily basis for vegetables and sometimes for other foods like meat or beans, an overall savings of 20 percent on cooking costs could be expected. A pressure cooker costs about $45.

Crockpots offer low-heat cooking for meals that can be prepared slowly, such as stews and roasts. They can be turned on before leaving the house in the morning and by dinnertime your food will be ready to eat. A full day's cooking typically runs about 9¢, which is a savings of at least 65 percent over meals cooked conventionally on or in your range. A Crockpot costs about $25.

The *wok* is ideally suited to gas cooking, less so for use with an electric range. A metal collar can, however, make wok use practical on electric stove tops. The wok is used to stir-fry, steam, or even deep-fry vegetables and meat chunks. Food cooks quickly at a high temperature. This approach boosts food quality by minimizing nutritional losses and can save about 80 percent on the energy required to cook a specific item. A wok with collar and lid costs about $30.

A *toaster oven* uses electricity, and generally speaking I do not recommend it. However, if you are committed to electric cooking, you can reduce your electricity consumption somewhat by cooking steaks, chops, or hamburgers in a small toaster oven instead of a range. Broiling in a toaster oven can provide a substantial savings over using the range's oven or broiler because it draws less electricity and is significantly smaller in size; a reduction of 50 percent per item cooked can be expected. A toaster oven costs about $75.

If you purchase all four of the above items (I especially recommend the pressure cooker and wok), you will spend about $175. If you use all four ap-

pliances regularly and also practice conservation when using your electric range, a savings of 50 to 60 percent over current cooking costs should be achievable. An annual outlay of $114 can be reduced to as little as $46, for example. Your investment in the appliances and in the conservation measures would be returned to you in three years.

Option 2: Convert to an Energy-Efficient Gas Range

The major advantage of shifting to gas cooking is the low price of natural gas relative to electricity. Going from an inefficient electric range to an energy-efficient gas model should reduce your bills by 65 to 70 percent. Unfortunately, not everyone can get natural gas. If you can get it but you do not already have a hookup, this involves added expense, but it is well worth it. Since the price of liquid propane (LP) gas can run quite high (up to about 80 percent the cost of electricity per usable Btu), it is a less desirable alternative.

Assuming natural gas is available to you, a $600 gas range will pay for itself in about six to seven years. Most of the conservation tips I gave earlier for electric cooking apply equally well to gas cooking; woks and pressure cookers are ideally suited to gas cooking. Thus, your 70 percent savings can be increased. In fact, an 85 percent decline in your current expenditure for cooking is possible.

Option 3: Convert to a Convection Oven

Because heated air naturally rises, a problem with conventional ovens is stratification (i.e., temperature differences between the upper and lower sections of the oven space). Convection ovens have two distinct advantages over standard models. The first is that they are smaller and thus there is less space to heat. The second is that they incorporate a small electric fan inside the oven in order to circulate the air and hence maintain a uniform temperature inside. (They derive their name from this feature: The heated air moves around inside the oven in a convection pattern.)

Convection ovens have a good track record as far as their ability to cook or bake, and they can save you 40 to 50 percent on electricity for these tasks. They are especially energy efficient for items that require an hour or more to cook. In some models, the fan may be moderately noisy, so you may want to compare several brands before you make a purchase. Convection ovens cost around $250 to $300 and should save you about $50 a year.

Option 4: Convert to an Energy-Efficient Electric Range

New electric ranges do have some worthwhile energy-saving features such as better insulated windowless doors, flat burner elements for better contact with the bottoms of pots and pans, and smaller ovens.

"Energy Guide" labels found on other appliances such as refrigerators and water heaters are not put on electric ranges. This means you must do some comparative shopping for energy-efficient features on various models. A new energy-saving electric range could save you about 20 percent over an older poorly designed and poorly constructed model, but to purchase it is not really cost-effective since you would have to spend $500 to $600 to save $23 a year, and it would take 20 years to get your money back. Options 1 and 2 remain more rational choices than getting a new electric range unless your old range is in poor condition and you are determined to keep cooking electrically.

Several new models use a magnetic induction cooktop that relies on electromagnetic energy to directly heat a pot. These ranges are relatively expensive, however ($2,000), so although they can save you 25 percent on cooking, they are probably not worth the investment.

Option 5: Convert to Microwave Cooking

The easiest way to save on energy used for cooking is to convert to a microwave oven. You will probably save anywhere from 40 to 75 percent a year. At a cost of $500 to $600, your investment would be returned in about seven to eight years. Note, however, that microwave ovens still exhibit a payback period three times longer than option 1 plus conservation. Although microwave ovens are undoubtedly convenient, they are not nearly as cost-effective as option 1. Secondly, there is the issue of safety. I personally avoid being around microwave ovens, even occasionally. To operate one in your own home is, in my own opinion, risky business. Some portion of the radiation in the unit is known to leak out to the surrounding environment. This leakage is within U.S. government safety limits, but U.S. limits on microwaves are far less stringent than those established in some other nations. For example, the microwave ovens used in the United States are allowed to leak radiation that is substantially greater than the levels considered unsafe in the Soviet Union. Levels of microwave radiation 40 times lower than U.S. standards have been shown to alter the brain chemistry of animals. This raises a question regarding the impact of long-term exposure to these devices. For this reason, I would not recommend microwave ovens. They can, however, save you a great deal of money.

Option 6: Use Solar Cookers

These devices are now available through several companies. They can be used for supplemental cooking, especially during the summer months. Essentially, they are solar collectors that use sunlight to heat your food. They are generally used outdoors.

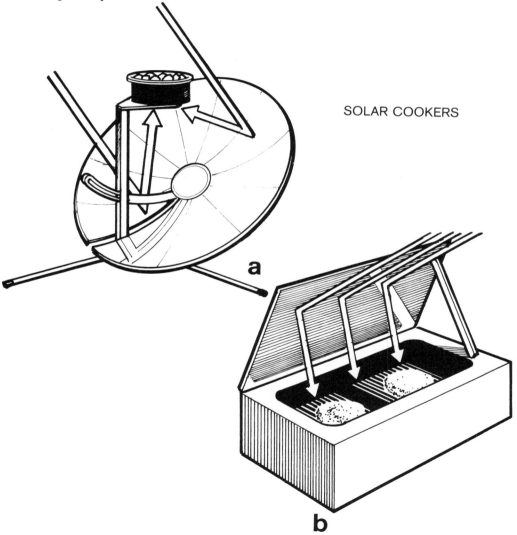

SOLAR COOKERS

Figure 5-1. The reflector of a focusing solar cooker (a) concentrates the sun's rays to heat a utensil containing food. The reflector of an oven-type solar cooker (b) directs the sun's heat into an insulated box.

There are two basic types of solar cookers. A focusing unit uses reflectors to concentrate the sun's energy so that high temperatures can be attained in a small area. This approach is similar to the effect of using a stove-top burner. A black heat-absorbing cast-iron pot or pan is ideally suited for this type of solar unit.

The other cooker acts more like an oven in that it uses mirrors or reflective foil to redirect sunlight into an insulated box. The food to be prepared is placed in the box; sunlight heats the air in the box to a high temperature. Many foods, such as roasts and eggs, can be prepared this way, and you can also bake bread.

The main disadvantage of solar cookers is that they can not be used on cloudy days or at night. The devices must also be reoriented toward the moving sun every 20 minutes or so. Despite the fact that solar cookers use free energy, they are probably not practical as a primary cooking method for most American families. At the present time, they are mainly used in the Third World, where cooking fuels are scarce.

Recommendations

1 Follow the 15 conservation steps and convert to supplemental cooking devices to help reduce the amount of electricity your range uses. The savings will be 50 to 60 percent with a three-year payback period.

2 Convert to an energy-efficient, natural gas range with a pilotless (electronic) ignition system and other energy-saving features. Follow the 15 conservation steps. Buy a wok and pressure cooker. Save 80 percent and repay your investment in about six or seven years.

For more information, see the section of part 2 titled "Ranges and Ovens."

Chapter 6
Cooling

If you currently use central or room air conditioning, you are spending a great deal of money to stay cool. Depending on where you live, you may need to cool your home from 0 up to 2,000 hours a year. Many people do little or nothing and suffer the heat; others air condition and suffer the cost.

The actual cost of staying cool with air conditioning in a given section of the country varies according to the size of your house and the amount of insulation it contains. In most northern areas, an 1,800-square-foot home can usually be cooled by an air conditioner with a 50,000-Btu capacity. A poorly insulated home needs an air conditioner with twice as much capacity, while a well-insulated home should need a unit rated at only 30,000 to 35,000 Btu. (Notice that cooling equipment, like heating equipment, is rated in British thermal units. The Btu capacity of an air conditioner describes its ability to remove unwanted heat. Thus the higher the Btu rating, the more cooling it can provide to a given space or the more space it can cool.)

The cost of cooling also varies with the efficiency of your equipment. New air conditioners are now labeled with an Energy Efficiency Rating (EER). This label lets you calculate how much electricity an air conditioner will consume. For example, if a 50,000-Btu central air conditioner has an EER of 8, then you know it will require 6.25 kwh (50,000 divided by 8) to give you 50,000 Btu of cooling power for 1 hour. Assuming an electric rate of 9¢ per kwh, the hourly cost to cool your home would be 56¢. An EER of 12 would allow you to meet the same cooling load for about 38¢ per hour. Thus, the higher the EER, the more economical your equipment is to operate.

Room-size air conditioners cost less to operate than central units because cooling is "zoned," meaning that you cool only one or two rooms, not the entire house. A 10,000-Btu room air conditioner with an EER of 8 would require 1.25 kwh, so at 9¢ per kwh, it would cost $101 a year to operate in an area of the country requiring 900 hours of cooling each year. Two room-size units would cost $202 to operate, while a 50,000-Btu central unit would cost about $500 to operate for the season.

Basic Conservation Measures

No matter what kind of air conditioner you may have, if you wish to reduce cooling costs, here are some tips on the way to do it.

Step 1: Tighten Up Your House

If you now air condition your home, you can reduce costs by up to 67 percent with a conservation retrofit. Insulation, caulk, and weather stripping can reduce your cooling costs. Table 6-1 indicates how increasing your home's insulation can bring down your annual cooling bill.

As table 6-1 shows, substantial reductions in air conditioning costs are possible by adding insulation to your home. Just as insulation holds heat indoors during winter, it keeps heat outdoors during the summer. The conservation retrofit that can achieve these savings is outlined in chapter 7. Tightening up your house with insulation, weather stripping, and caulk is very cost-effective, even if you do not heat or cool electrically.

Table 6-1

Annual Cost to Air Condition 1,800-Square-Foot House*

Insulation	Atlanta (7¢ per kwh)†	Dallas (9¢ per kwh)	New York (15¢ per kwh)
Poor	$819	$1,458	$1,013
Moderate	$573	$1,122	$709
Excellent	$307	$547	$380

*The air conditioner has an EER of 8.
†The rates shown are local averages in the three cities.

Step 2: Make Your Existing Air Conditioner More Efficient

The hotter it gets, the less efficient your air conditioner gets. By modifying the environment around your air conditioner to make it cooler, you can increase the unit's efficiency. This means you receive more cooling for your money.

If you keep the sun off your air conditioner, either with a shading device such as an awning or with trees and other vegetation, you will boost its efficiency. A thick tree canopy can block 70 percent of the sunlight and reduce surrounding temperatures by 10°F to 15°F. An additional advantage is that plants generate cooling moisture through their life processes. This also benefits your air conditioner.

Step 3: Turn Up Your Thermostat

People often set the thermostats on their air conditioners too low. The temperature inside your house or in a cooled room does not have to go below 78°F. Your air conditioner not only cools but also dehumidifies, and this stimulates evaporative cooling of your skin. This makes you feel cool even when the temperature in the room is fairly high. Think of how often you have felt chilled in an air conditioned room. If you experienced the same feeling in winter, you would probably turn up the heat.

The impact of turning up the temperature setting on your air conditioner from 72°F to 78°F is phenomenal. This simple conservation measure can save families with central air conditioners hundreds of dollars a year. The cost is zero and the payback is immediate. In the hottest areas of the nation (i.e., southern Florida and the Southwest), you can save a minimum of 15 percent on your annual cooling costs. In the remainder of the South, the savings are around 20 percent, and in the rest of the nation, you may achieve a 25- to 30-percent drop in your cooling bill.

Step 4: Zone and Time Your Equipment's Operation

If you have central air conditioning, you waste a great deal of energy cooling space that is rarely or never used. Closing off unused areas can reduce your cooling load. A room air conditioner is meant for "zone" cooling of this sort, so make sure doors are kept closed in rooms with such air conditioners.

You should consider replacing your air conditioner's existing thermostat with a timing control or think about buying an attachable timer. This will turn your air conditioner off in the cooler early-morning hours and during periods of the day when no one will be at home. You could save 10 to 15 percent of your annual cooling costs this way. Of course, the cheapest option of all is to train yourself to turn off the air conditioner manually during the hours when it isn't needed.

Step 5: Purchase an Automatic Misting Device

Precooling misters are used to bathe the condenser coils of a central air conditioner in a fine spray of water. This cools the unit down and makes it run more efficiently. An annual reduction of 20 percent is typical. Fully automatic devices with a sensing capacity sell for $150 to $350 and should pay for themselves in two to three seasons. A simpler device connects to your garden hose and provides a continuous spray of ½ gallon of water per hour. It costs $15 and has a payback period of several weeks during the cooling season.

AIR CONDITIONER MISTER

central air conditioner

mister

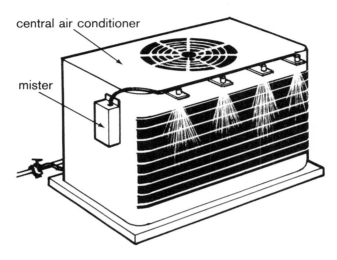

Figure 6-1. An air conditioner may become 20 percent more efficient if it is equipped with a misting device to cool the air conditioner's coils.

Misters can be cost-effective investments in terms of energy savings, but there are several problems associated with their use. Bathing an air conditioner's metal parts in water on an ongoing basis can cause mineral scale to build up, especially in areas with hard water. A mild acidic solution (available from mister manufacturers) can be used to periodically remove the mineral buildup. A second problem is corrosion of the metal itself. This problem is not confined to air conditioners with misters. Normal moisture buildup due to condensation and rain will have the same impact. One mister manufacturer recommends spraying coils and fins with silicone to seal them off. A third, minor problem is that misters consume water in the process of saving you electricity. If you spray your air conditioner for 16 hours a day, you consume an extra 8 gallons of water. This is a relatively small quantity of water based on current American consumption levels.

For more information about misting devices, see the section of part 2 titled "Energy-Efficient Air Conditioners."

Other Options

The above conservation measures can make a major dent in air conditioning costs. The following options also hold out varying degrees of cooling potential:

Option 1: Convert to More Efficient Room Air Conditioning

If you have an older central air conditioner, consider switching to several high-efficiency room units; you will experience a monumental drop in your electric use. If you already use room air conditioners, you will find that newer models with an EER of 10 or more can save you a substantial amount of money over the long run. This is especially true if you live in a warm climate or where electric rates are high.

Option 2: Convert to More Efficient Central Air Conditioning

An EER of 5 gives you half as much cooling per dollar as a rating of 10. The range of EERs is 5 to 13. If you have an old air conditioner, it's probably at the low end of the scale and could cost you a small fortune every season to operate. With an air conditioner rated at EER 12, you could pay $750 and get the same cooling that you would from a model rated at EER 6 and costing $1,500 to operate. The greater the discrepancy between your air conditioner and the most efficient new models, the more logical it is to get new equipment.

A study in Florida shows an average family there saves 33 percent on air conditioning costs after converting to new, more energy-efficient equipment. Newer units have a larger coil surface to promote more efficient removal of heat and better control systems that boost compressor operating efficiency. When you purchase a new air conditioner, look for one with an EER of 10 or higher. Research has shown that if, during the month with the highest electric bill of the summer, you use twice as much electricity as you do during the month with the lowest electric bill, you are probably better off replacing your present model sooner rather than later. I personally believe central air conditioning is inappropriate since so much unused space gets cooled. I would recommend you pursue option 1 before deciding to continue with a central system.

Option 3: Get a Reverse-Cycle Air Conditioner

If you now have electric space heating, you might want to consider one or several high-efficiency air conditioner/heat pump units. During the spring and fall, they provide heat at about 40 to 50 percent the cost of conventional electric heating. In summer, they run in reverse and extract heat from inside your home and send it outside (i.e., they air condition). Winter temperature extremes curtail their efficiencies, but their overall impact on yearly energy costs can be substantial. Costing $600 to $850 to buy, they should pay for themselves in five to six years, thanks to the savings on electric heat alone. If you replace a relatively inefficient air conditioner with an efficient reverse-cycle unit, your savings will be even greater since your costs will go down in both cold and warm seasons.

For more information about air conditioning, see the section of part 2 titled "Energy-Efficient Air Conditioners."

Option 4: Use Passive/Low-Energy Techniques

The best thing to do with an air conditioner is to use it infrequently—or, if possible, not at all. Various alternative cooling techniques are possible.

To understand these techniques, a little background information is helpful. A home's cooling load is determined by the size of the home and its ability to resist the movement of heat from outdoors to indoors. Clearly, the hotter the area where you live, the harder it is for your house to block out the heat. As table 6-2 indicates, there is a substantial difference in average air temperatures within the United States.

Table 6-2

Average Summer Temperatures in Selected U.S. Cities

City	Temperature (°F)	City	Temperature (°F)
Albuquerque, N. Mex.	80.8	Las Vegas, Nev.	92.0
Atlanta, Ga.	81.6	Lexington, Ky.	78.0
Boise, Idaho	74.2	Lincoln, Neb.	79.6
Boston, Mass.	73.6	Little Rock, Ark.	82.1
Charleston, S.C.	82.0	Los Angeles, Calif.	74.2
Cleveland, Ohio	74.9	Miami, Fla.	83.8
Columbia, Mo.	78.8	Nashville, Tenn.	81.7
Dodge City, Kans.	81.1	New York, N.Y.	74.8
East Lansing, Mich.	72.4	Oklahoma City, Okla.	83.8
El Paso, Tex.	84.2	Phoenix, Ariz.	92.1
Fort Worth, Tex.	86.8	Salt Lake City, Utah	77.3
Fresno, Calif.	84.4	San Antonio, Tex.	86.7
Greensboro, N.C.	79.0	Tampa, Fla.	83.8
Indianapolis, Ind.	77.3	Tucson, Ariz.	88.2
Lake Charles, La.	84.4	Washington, D.C.	78.0

Locations in the Southwest, south Texas, and south Florida have the highest cooling requirements, between 1,500 and 2,000 hours per year. The remainder of the South and the lower Midwest require 1,000 to 1,500 hours of cooling annually. The mid-Atlantic states, the Midwest (excluding northern sections), the western mountain states, and the lower coast of California average between 500 and 1,000 hours annual cooling need. The remainder of the nation has a cooling requirement of 500 hours a year or less.

Air temperature is not the only determinant of human comfort. Humidity, air flow, and the radiant temperatures of wall, floor, and ceiling surfaces are also important. In the southeastern section of the nation, for instance, average summer temperatures of 80°F to 85°F may "feel" less comfortable than higher temperatures of 85°F to 90°F in the less humid Southwest. Air conditioning aims primarily at reducing temperatures in your living space and dehumidifying the air. The techniques discussed below can reduce interior air temperature and influence other comfort factors at only a fraction of the cost to air condition.

Technique 1: Insulate Your Home

As I mentioned earlier, insulation can be just as important in hot weather as it is in cold weather. The most important area to insulate is the attic, since heat can pour down into the house from the roof, which is bombarded by sunlight. Specific suggestions concerning insulation are given in the next chapter.

While you insulate your attic, you should also consider installing perforated aluminum foil in the attic. Specifically, you can staple the foil to the bottoms of your rafters, so that the foil forms a continuous surface stretching between the rafters. Studies indicate that perforated aluminium foil stapled to the attic rafters can do an excellent job of reducing radiant heat flowing from your roof to your living space. The heat radiating into your attic from a sun-baked roof will be blocked by the foil.

Technique 2: Install Shading Devices

The amount of heat entering your home through windows can be reduced by 80 percent with awnings, overhangs, and shutters. An outside temperature of 75°F can mean an inside temperature of 90°F when direct sunlight penetrates your home. By simply blocking this sunlight, you can shave 25 percent from your annual air conditioning bills. It is the logical second step (after insulating your attic) toward lowering temperatures to a level where air conditioning will rarely be needed.

Shutters and awnings mounted outside do a good job of blocking the sunlight. Shading devices mounted on the inside of windows provide some reduction, but part of the sun's heat does get into your living space. The advantage of interior drapes, blinds, or shades, however, is their convenience relative to exterior devices.

Some homes are now built with permanent exterior overhangs. These are ideal in that they block the summer sun, which is high in the sky, but not the winter sun, which strikes your home at a lower angle. These overhangs are used on the south side of the house. South-side awnings can serve a similar purpose. The east and west sides of a house receive morning and late afternoon sunlight at a low angle even in the summer; thus overhangs are generally ineffective on these locations. Instead, exterior shading devices are recommended to block east-side and especially west-side sunlight.

A shading retrofit to block sunlight will vary in cost. A combination of shading devices on the east, south, and west sides can be installed for as little as $250 to $300. Awnings will make this package somewhat more expensive. If you now use central air conditioning, you could save $100 to $150 per year. Your payback period would be two or three years.

Another sun-blocking option is the installation of a transparent, metallized window film. Window films are sheets of treated plastic that reduce the amount of light entering a window while still letting you see through the window. They are relatively inexpensive—50¢ to $3 per square foot—and will reduce incoming solar radiation by up to 80 percent. If you also need the sun's warmth in winter, make sure you purchase a film that can be removed in winter and reinstalled during the warmer season. Solar screens perform a similar function. Available in major department stores, they are like insect screens but with a mesh designed to intercept sunlight. Placing such screens on your windows can reduce the inside daily temperature of your home by 10°F or more.

Yet another possibility is the use of a shade wall as a sun-blocking device. The wall itself is constructed of masonry block about half the thickness of ordinary cinder block (about 4 inches thick) and contains open air vents to allow air flow through the wall. Painted white to reflect the sun's heat, the wall is usually constructed parallel to the west wall of a dwelling so that it blocks the low afternoon sun. Because it is placed 3 to 4 feet out from the house and is open both at the front and back and via the air vents, air movement is sufficient to provide ventilation but direct sunlight is virtually eliminated. This passive technique is common in Australia and is used to a lesser extent in the southern United States.

Technique 3: Use Vegetation to Cool Your Home

Plant trees, vines, and hedges around your home. On the south side, make sure trees are deciduous so they will lose their leaves in winter and allow sunlight through. In summer, trees can block up to 70 percent of incoming sunlight. Vegetation also creates a cool, moist climate near your house. Incoming air is cooled by the vegetation and shade before it enters your home, and this keeps interior temperatures more comfortable.

The trellis is an ideal method to use in conjunction with vegetation to reduce incoming sunlight. Plant vines and allow them to climb over a trellis positioned

to shield one or more windows from the sun. The vines will grow much more quickly than trees and you will be using Mother Nature to help cut your cooling costs.

Technique 4: Get the Air Inside Your House Moving

As I have pointed out, cooling is not only a matter of lowering the temperature. Humidity and the movement of air also are important to how you feel. If 82°F air is blowing at 5 miles per hour (a gentle breeze), you will feel as comfortable as you would in 72°F air that is motionless. Air movement can be facilitated either naturally or artificially.

Natural convection is the simple process by which warm air rises. Any opening or vent at a high point in your living space such as an open skylight or high window will act as a natural exit for hot air. When hot air leaves your house, it will be replaced by air from outside. If this incoming air first passes through vegetation or over water, or if it comes from the north side of your house or up from the basement, it will cool you more than if it comes from hot places (e.g., from unshaded areas, over roads, or the south side of your house). By using high vents on the south and west and low vents or windows on the north side of your home, you will create natural interior breezes that will make you feel comfortable without the aid of air conditioning.

In addition to the high vent or open skylight, you may want to consider a thermal chimney. This is a chimney—often made of sheet metal—extending through your roof or running up one of your home's exterior walls. Sunlight heats the chimney, so the heated air in the chimney rises and exits through louvers at the top. This causes air from inside the house to rise into the chimney, replacing the air that has left. In turn, air from the basement or from outdoors is then drawn into the house to replace the air that has risen into the chimney. The chimney should be situated so as to draw air from a cool area of a house (e.g., the basement or crawl space) through the house's living spaces. The best way to retrieve this cooler air is by providing a direct path for it, an open stairwell for instance. The chimney is glazed on the west side and partially on the south so as to further warm the air in the chimney and thus speed up its natural flow upward. A wind turbine can also be mounted atop the chimney. The wind drives the turbine (makes it turn) and this also helps to pull air out of the chimney.

Fans provide an alternative means of moving air at a fraction of the electrical cost of air conditioning. You should be careful about where you position your fans, however. Ceiling-mounted paddle fans, for example, may be counterproductive in that they take the warmest air—at the ceiling level—and send it down to you.

Attic fans (fans designed to cool the attic) have little effect in cooling the rooms below the attic. Mounted in the attic gable walls, they vent hot attic air to the outside. This may be beneficial to the attic itself, but it won't do much to cool the rooms below the attic. Research has shown that with even a few inches

THERMAL CHIMNEY

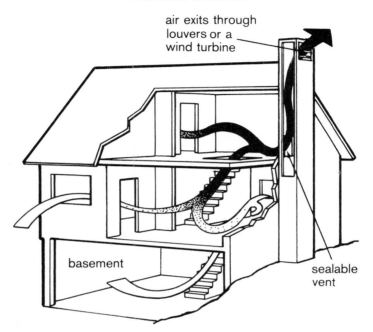

air exits through
louvers or a
wind turbine

basement

sealable
vent

Figure 6-2. A thermal chimney promotes natural convection throughout a house:
Heated air rises out of the chimney, pulling cooler air through the house's living
spaces.

of insulation in your attic, very little heat enters the house from the attic, so the
small reduction in air conditioning costs that results from cooling the attic is
generally offset by the operating expense of running an attic fan. (Note that attic
fans become even less necessary if you have stapled perforated aluminum foil
to your rafters, thus helping to keep the attic cool.)

Whole-house fans seem to be the most desirable type of fan. A whole-
house fan is a large fan, recessed in a ceiling, that expels warm air from the
house into the attic while pulling cooler outside air into the house from the
basement or through first-floor windows. Such fans operate at 10 percent of
the cost of air conditioning, yet they are highly effective. The heat sent into the
attic by a whole-house fan is vented to the outside through gable vents. Without
adequate insulation directly over your ceiling, however, you would defeat the
purpose of the whole-house fan, since much of the heat in the attic would just
return to your living space. So once again the importance of adequate insulation
becomes apparent.

The operation of a whole-house fan is simple. Houses warm up as the day
goes on—about 1°F each hour. Outside temperatures begin to drop about 6:00

WHOLE-HOUSE FAN

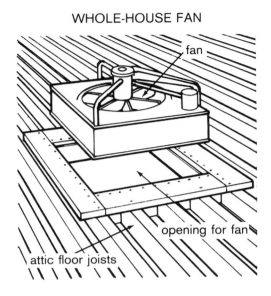

Figure 6-3. A whole-house fan is installed in a ceiling: It pulls warm air out of the house and expels it into the attic.

P.M., but houses retain heat, so they cool off less quickly. A whole-house fan does two things. It creates a breeze within the house that makes you feel 4°F to 10°F cooler than it actually is. It also exchanges interior air for outside air at a rapid rate—about 40 to 60 times per hour.

Consider a hot day. At 6:00 A.M., your house is 70°F—this is comfortable. By 3:30 P.M., the outside temperature has reached 90°F. The temperature inside your home is somewhere in the low 80s. It is lower than the outside temperature because you have attic insulation and shading over the windows. As the inside temperature approaches an uncomfortable level, you turn on the whole-house fan. A breeze is created and even though the interior air temperature is above 80°F, you feel cool. If wise judgment is used, air will be vented high and to the south or west, and windows or vents on the cooler north side will provide fresh air. As outside temperatures drop in the evening, cooler air is constantly brought into contact with warm interior surfaces of the house. This cools them down. The impact of the fan is thus to generate a cooling "wind chill" factor and to rapidly facilitate the exchange of air. Forty exchanges per hour is adequate in most sections of the country.

The size of the whole-house fan you will need is determined by the size of your house. If you have 1,800 square feet of living space and your ceilings are 8 feet high, the volume of air inside is 1,800 multiplied by 8, or 14,400 cubic feet. To get 40 air exchanges per hour, you have to move this volume times 40 or 576,000 cubic feet per hour. Since fans are rated by the minute, you merely

divide the latter figure by 60 to reach a result of 9,600. Therefore, your fan will need a rating of about 9,600 cubic feet per minute (cfm). A unit this size costs about $500 installed or $250 to $300 if you install it yourself. You will find that a whole-house fan can reduce the need to air condition by 80 percent or more and can save you hundreds of dollars a year on electricity.

Whole-house fans may require a larger gable opening than you now have to insure proper venting of attic air. The fan blows warm indoor air into the attic; the air must then be able to escape from the attic to the outdoors. A 4-by-4-foot opening on both ends of your house is adequate for a 10,000-cfm fan, if you have wooden louvers. A 3-by-4-foot opening is adequate with metal louvers. A timer or manual switch can control the times of day your fan operates. A thermostat that responds to temperatures is not recommended. If a fire should occur, the fan would come on and generate an incredible breeze to feed the fire.

The smaller box-type window fan is also a suitable cooling device. Located high on the south or west side of your house, such a fan can evacuate hot air from rooms and allow cooler air to enter from elsewhere. Large powerful window fans costing about $150 can provide substantial air exchange without the installation problems associated with whole-house units. When you buy a fan, remember that the more air moved (cfm) for a given amount of watts used, the more efficient the fan is and the cheaper it is to run.

Technique 5: Modify the Impact of Interior Heat Sources

Dryers, lights, and cooking ranges generate substantial interior heat. Reduce the use of your oven in summer, use your dryer in the evening, and vent it to the outside. Better yet, use an outdoor clothesline in summer to dry clothes and an indoor one if possible in winter. The latter provides an ideal source of household moisture for households where space heating creates a problem of low humidity. It also saves you a great deal of money on both clothes drying and the operation of a humidifier. Cook meals later in the day. Follow the tips I gave earlier concerning refrigerators. Do not burn lights during the day—use natural light. Be particularly conscious of these tips on days that are especially hot.

Recommendations

⬜1⬜ Put insulation in your attic, walls, and floor to save on both cooling and heating losses. Install reflective foil between your rafters. Use sun-blocking devices—an awning on the south side, outside shutters or inside blinds or drapes on the east and west sides, or reflective film on all three sides. Plant vegetation around your house. Evergreens on the north side help provide cool summer air and can be an ideal winter windbreak. Poplars grow quickly and will help block

direct summer sunlight to the south, east, and west. Open north windows and install several small window fans or one large window fan on the south and west sides. If you have an air conditioner, use it only on those rare occasions when both temperature and humidity become unbearable.

2 If you feel you must air condition your home, cut operating time with supplemental shading and ventilation. Follow conservation tips. Run the dehumidifying setting on your air conditioner without cooling and see if this is sufficient on many occasions—it should make you feel more comfortable at a given temperature than when the humidity is higher. Cool specific rooms only during the hours when they are occupied. Bedrooms normally do not need daytime cooling and there is rarely a need to run your equipment after 1:00 or 2:00 P.M. If you have central air conditioning, use a mister to precool your condenser. If you buy a new air conditioner, make sure it has an EER of 10 or higher. If you want electric heat, a reverse-cycle heat pump unit may save you money on winter heating bills.

Chapter 7
Heating

Baseboard electrical heating systems are cheap to purchase and install. They are not cheap to operate, however. Many builders cut costs by installing electric heating systems in new homes, but the people who wind up using these systems also wind up with astronomical winter electric bills. If you're a victim of electric heat, you may have already taken some steps to reduce your heating costs. But if you have not already curtailed your power consumption—or if you have and wish to reduce it further—the following pages should help.

Conservation: The Essential First Steps

The amount of heat energy required to maintain your home at a comfortable level of air temperature throughout the heating season—let's assume 70°F— depends on several very basic factors. These are the climatic conditions where you live (i.e., outside air temperatures and prevailing wind speeds), the amount or volume of space to be heated, and the ability of your home to prevent heat from escaping to the outdoors.

You can do little to change climatic conditions, although planting windbreaks (rows of trees and/or bushes that block the wind) and orienting a house toward the south can help keep the house warm. Your home, assuming it is not still under construction, is fixed in terms of its size, although you might close off rooms and not use them or heat them in winter. The major path open to you for influencing your annual heat load is the third factor—the flow of heat energy out of your home.

Consider table 7-1. It shows the impact that improving your house with insulation, caulk, and weather stripping can have on reducing your expenditures for electric heating. The R-value of your insulation indicates how effectively heat is retained by the insulation. The higher the R-value, the better.

Table 7-1

Heating Costs for 1,500-Square-Foot House in Columbia, Missouri

Insulation Status	Description	Annual Cost at 8¢ per kwh*
Poor	Nothing done; R-1.7 ceiling; R-4.1 walls; R-3.1 floor; 1½-inch solid-wood door, single-pane windows, no caulking or weather stripping; 1.5 air changes per hour.	$6,000
Fair	Addition of 6 inches of fiberglass in ceiling and 6 inches of loose-fill insulation in walls; R-20.7 ceiling; R-19.3 walls; R-3.1 floors; storm windows, basic caulking and weather stripping; 1 air change per hour.	$2,265
Good	Fiberglass increased to 9½ inches in ceiling; 6 inches loose-fill in walls; 6 inches fiberglass added to floor; R-31.7 ceiling; R-19.3 walls; R-22.1 floor; insulating shades or shutters on windows; proper caulking and weather stripping; .5 air change per hour.	$1,158
Excellent	Same as "Good" home except ceiling has 12 inches of fiberglass (R-39.7).	$984

---

*8¢ per kwh represents a 1¢ discount from the average price of 9¢ per kwh. Many electric companies offer a discount for customers who use electric heat.

Step 1: Insulate Above Your Ceiling

If the example home described in table 7-1 has little or no insulation ("poor" insulation status), heat from inside passes quickly to the outside environment. Thus, this heat must be replaced by more heat. In this outrageous but possible situation, the annual cost of heating with electricity would be $6,000. Close to one-half of this outlay is due to heat being lost through the uninsulated ceiling. Obviously, installing ceiling insulation is the first priority in any conservation retrofit project.

Our example home is just one story tall, and it has a 1,500-square-foot ceiling area. Installing 12 inches of fiberglass insulation in the attic would cost about $800. A two-story home of the same size would have 750 square feet of

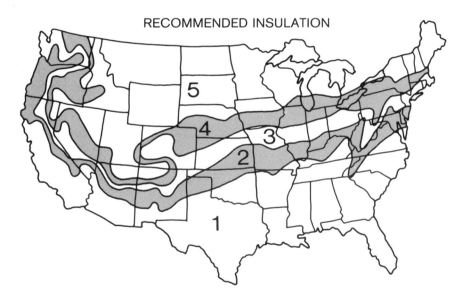

RECOMMENDED INSULATION

Figure 7-1. For maximum energy efficiency, homes should be insulated to the values shown.

Region	Attics and Ceilings	Walls	Floors over Unheated Areas	Foundations and Below-Grade Walls
1	R-26	R-12	R-12	R-8
2	R-38	R-20	R-16	R-12
3	R-44	R-26	R-20	R-14
4	R-50	R-32	R-24	R-16
5	R-60	R-40	R-30	R-20

second-floor ceiling and 750 square feet of first-floor ceiling. Only the second-floor ceiling would need insulation (since it is the only ceiling through which heat can escape to the outdoors), so the cost of 12-inch fiberglass for this ceiling would be about $400.

Twelve inches of fiberglass would increase the R-value of a previously uninsulated ceiling from 1.7 up to 39.7—an increase of over 2,300 percent. It would have the effect of reducing annual costs attributable to ceiling heat losses from $2,635 down to only $110—a 96 percent reduction! With a one-time investment of $800 in insulating material, the savings to you during the first year would be $1,725, and during each year thereafter the savings would be over $2,500.

There are numerous other insulating materials besides fiberglass such as mineral wool, extruded polystyrene, polyisocyanurate, cellulose, and, of course, the notorious urea formaldehyde, which is not recommended. The costs of these materials vary, with fiberglass at the low end of the cost range. No matter which type of insulating material you choose, each is a bargain if you have electric heat. Since fiberglass is cheapest and, once it is in your attic, constitutes a lower health and fire hazard than most other insulations, it is the one I recommend.

You should be aware that, beyond a certain point, adding more and more insulation is not advisable. There are diminishing returns in savings relative to your investment as you pile more R-value into your attic. For example, in the poorly insulated house, adding 3½ inches of fiberglass would provide a whopping 87 percent savings on ceiling heat losses, but there is not all that much left to save as you add additional amounts. If you went from no insulation to 6 inches, your heat loss reduction would be 92 percent—only 5 percent more than with 3½ inches. Twelve inches would save you 96 percent over no fiberglass, but you've increased the R-value threefold by adding 12 inches instead of 3½ inches and yet you've only achieved an extra 9 percent savings on annual heating losses and costs. Since electric heat is very expensive, however, a 9 percent incremental savings is worth pursuing—at our example home it would put another $237 in your pocket each year. My recommendation is that you put 12 inches of fiberglass or the equivalent in your attic. You can add more if you wish, but at the current cost of electricity, it is not economically wise. At higher future rates, it may be worth doing.

Step 2: Insulate Your Walls

People often believe that once you've insulated your ceiling, you've won the war on heat loss, so insulating the walls is not particularly important. It's true that walls do not lose heat as quickly as ceilings do, but there is room for substantial savings here, nonetheless.

In our 1,500-square-foot example house, there are 938 square feet of exterior walls. (The home measures 30 by 50 feet, with 8-foot-high ceilings, 300 square feet of windows, and 42 square feet of doors.) The annual heat loss through the walls (not including windows and doors) is equal to $670. The addition of 6 inches of loose-fill insulation to these walls would reduce the wall heat losses 79 percent by increasing the average R-value from 4.1 to 19.3. This would save $530 per year. Many older homes are framed with 2 × 4 studs, and thus cannot accommodate as much insulation in their walls. Still, insulating the walls would significantly lower these homes' heating bills.

Blown-in loose-fill insulation costs about 60¢ per square foot. The once-only investment at our example home would run about $560. This gives a payback period of just slightly more than one year. At higher electric rates, the payback period would, of course, be even less.

Step 3: Insulate Under Your Floor

If your living space is over an unheated basement or crawl space, and if the area under your floor is not insulated, your annual heat losses through the floor are probably substantial. You can expect to lose about half the heat energy per square foot through the floor that you lose through walls. Many people mistakenly feel that wall-to-wall carpeting with a rubber undermat sufficiently reduces cold at the floor level. Such a carpet, placed over ¾-inch hardwood and 2-by-8-inch joists, provides only about R-3 in insulating value. The seasonal heat losses through an R-3 floor for a 1,500 square foot single-story house with an average basement or crawl space winter temperature of 53°F is about $722. The addition of 3½ inches of fiberglass (foil toward the interior of your house) can save $558 of this outlay. This is a 77 percent savings. Five and a half inches of fiberglass would lower the heat losses by 84 percent, bringing the cost of heat lost through the floor to $116 a year. A $600 investment in 6-inch fiberglass insulation would save you $606 per year, in these circumstances.

Many homes are built on a concrete slab. Because cement has a very poor insulating ability and because the earth itself is always cooler in winter than the temperature of a comfortably heated house, losses through the floor occur. The installation of 1½-inch extruded polystyrene board around the perimeter of the slab to 24 inches below grade would reduce floor heat losses by 50 percent.

If you heat your basement, you should insulate it. Installing 1½ inches of extruded polystyrene to 24 inches below grade around the outside of the basement walls will provide an overall heat-loss barrier of R-10 to R-12, depending on the extent of foundation wall exposure. Alternately, you can install 2 × 4 studs and 3½ inches of fiberglass on the inside surface of the basement walls, providing an overall insulation value of R-13 to R-16, depending on the amount of above-grade outside foundation wall exposure. This technique should run about $1 per square foot.

Step 4: Install Storm Windows and Insulating Shutters or Drapes

The average heat loss through a window is influenced not only by outside temperatures and the R-value of the glazing but also by the orientation of the window and the amount of sunshine that strikes it. South-facing windows have a lower net heat loss due to the solar warmth they receive during sunny periods. North-facing windows, on the other hand, are major heat losers since the sun never shines directly on them. East- and west-facing windows fall somewhere in between north- and south-facing windows in their net heat loss.

Well-designed energy-efficient homes emphasize south glazing and minimize east, west, and especially north glazing. Most American homes are not designed to take advantage of the natural heat of the sun. Our example home with 300 square feet of single-pane windows having an R-value of approximately 1 would lose $983 worth of electric heat through the windows each year. Installing R-4 insulating shutters or drapes and storm windows should reduce heat losses through the windows by 71 percent. The best insulating devices are interior shades, quilts, and shutters, especially those that form a tight seal over the windows. At $50 per 3-by-5-foot window for storm windows and $60 and up for insulating shades, a full retrofit would cost well over $2,000. Despite this ominous figure, window losses would be reduced from about $980 to $280 worth of kwh a year. The payback period would be only three years, based on a savings of $700 per year.

You may wish to consider installing a third layer of glazing as an alternative to insulating shutters, shades, or drapes. (This would be an additional layer added to double-glazed windows or to single-glazed windows that are equipped with storm windows.) If you have double-glazed windows, Plexiglas panels that fit snugly inside your window frames can offer almost as great a reduction in heat loss as insulating shades because, whereas the shades are only closed at night, the third layer of window glazing is always in place. Triple glazing is most appropriate for north windows, since no solar heat is gained through windows on this side of your home, so the extra glazing will not reduce the amount of heat you receive from the sun. In any case, window inserts should always be removable in summer so as to allow cool outdoor air to enter your house.

Step 5: Weather Stripping and Caulk

Insulation and window covers inhibit one type of heat loss: heat that is conducted through your home's outer surfaces. A second way you can lose valuable heat is through infiltration—the influx of cold outside air into your home and the simultaneous escape of warm inside air. This occurs wherever seams, cracks, and holes exist in your house. Typically, air leaks are found between the

foundation and framing sills, where pipes or vents enter or leave your house, and most especially around the frames of windows and doors. Our example house would lose 36 percent of its heat this way ($956 worth of electric heat) if it were very leaky. Proper caulking and weather stripping could cut these losses by two-thirds.

Weather stripping provides a seal for movable windows and doors. Because both windows and doors must be able to open and close freely, a completely tight fit between their surfaces is impossible. Weather stripping—long strips of material placed inside door and window frames—compensates for this by blocking air movement when windows and doors are closed. Many weather strippings are not adequate to their task, however. Some actually make air leaks worse. The most common types of weather stripping are compression gaskets made of felt, foam, and sponge rubber. They all operate by being squeezed between two surfaces (such as a door and its frame) in order to create an airtight seal. Unfortunately these materials tend to disintegrate in a short time from use, and this limits their value.

Tension strips are long sections of thin metal or plastic with a V or L cross section. They are springy and thus place continuous pressure on doors and windows. These generally tend to be a better choice than compression weather strippings. Door bottoms are best sealed with a "sweep." This is a flap of rubber or other materials that prevents air from flowing through the crack between the bottom of the door and the floor. Weather-strip all doors and windows in your house.

Caulk is a thick, viscous material that is used to plug gaps in the surface of a home. Typically, caulk is applied to the seams where the outer window and door frames meet the exterior siding of your house, where the foundation meets the sill plate, and where any gaps exist around pipes, vents, or electrical connections. Many types of caulk are available. The caulk you use should contain a silicone base. This assures that it will expand and contract with your house as the weather changes. Inflexible caulks will dry and crack, allowing cold air to leak into the house. Most houses need up to a case of caulking tubes, so you might save money by purchasing a full case.

Electrical outlets often permit infiltration: They are, in effect, holes in your walls. If you place your hand in front of an exterior wall outlet and feel cool air, you should install a foam rubber gasket behind the cover plate. These gaskets are available at hardware stores. Turn off the electrical power, remove the cover plate, install the gasket, and replace the cover again. Do this at all the exterior-wall outlets. It should take 30 minutes to do your entire house.

A kitchen exhaust fan usually is a huge heat loser. Close it off with insulation and removable caulk during the cold months. Pet doors and mail slots are equally bad. If your cat has to go out, open the door for him or her (quickly) and close off the pet entrance for the season. Seal mail slots and put a box outside.

If you have a fireplace, be sure to keep the flue damper closed when the fireplace is not in use. Fireplaces are an ideal way to stimulate air currents in your house and to send warm air to the sky above. If your home's central heating

WHERE TO CAULK

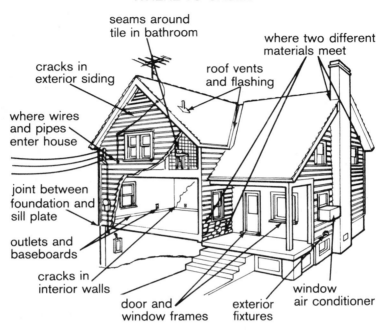

Figure 7-2. To help prevent cold outdoor air from infiltrating the home, caulk should be applied in all of the areas shown.

system is on, you may actually experience a 10 percent greater heat loss with the fireplace going than without it: The fireplace sucks in air that your heating system has warmed, and it shoots it up the chimney. This means that the overall efficiency of your fireplace is negative. You are worse off if you use it than if you don't.

If your house is well constructed and has some caulk and weather stripping already, your infiltration is probably equal to one air change every hour. With electric heat at our example home, we would lose $643 per season this way, assuming an electric rate of 9¢ per kwh. By using the right weather stripping and the proper caulk for each job, and by attacking all possible leaky areas, we could reduce infiltration to half an air exchange per hour. The total cost to caulk and weather-strip would be $200 to $300 and the payback period would be about one year.

Step 6: Set Back Your Thermostat

The real issue in heating is not temperature but comfort. When you insulate your house, you also reduce the coolness of interior surfaces. When you tighten

your house with caulk and weather stripping, you eliminate drafts. Cold surfaces and moving air make you feel cold. Warm the surfaces and reduce the movement of air, and you will feel comfortable at a lower temperature. After a conservation retrofit, 68°F or 69°F should feel fine to you. A 6°F reduction in the average temperature of your home may save you 12 to 15 percent on fuel bills. A 10°F nighttime reduction could do the same thing.

The most convenient and logical approach to temperature setbacks is the automatic or clock thermostat. For $60 to $90, you can buy a quality unit that will allow two or three daily setback possibilities—that is, it will automatically change the thermostat setting at two or three predetermined times during each 24-hour period. For less money, you can buy a manual device that allows you to lower your thermostat's setting each night and elect a time for the thermostat to bring the heat back up again the next morning. Many excellent thermostat-control devices are available.

Wise use of thermostat timing equipment (i.e., a 15°F nighttime setback and a daytime reduction of a few degrees) can bring you an additional 20 percent savings over those provided by steps 1-5 plus step 8. This would reduce the annual electric heating bill at our example home to about $787.

Step 7: Redistribute Your Heat

Homes with ceilings 10 to 16 feet high may experience problems of heat stratification (i.e., the heat "stratifies" by rising to the ceiling, leaving cooler air near the floor). Heat naturally rises, and the temperature differential between ceiling and floor levels can be as much as 15°F. The air can be a cool 60°F where you are while an untapped source of 80°F warmth lies only a few feet away.

If you can capture upper-level heat and mix it with the air below, you can keep your thermostat a bit lower and thus save energy. The most popular device for redistributing ceiling heat is the destratifier. This is a device that gently pulls warm air from the upper levels of a room and sends it out at floor level. It establishes a more equitable mixing of air currents and thus of air temperatures at the ceiling and floor levels. It costs about $40 and can save you up to 12 percent on your heating costs. It is marketed through energy stores and through popular energy-oriented magazines such as *New Shelter* and *Solar Age*.

Step 8: Install Storm Doors or Insulating Doors

A typical 1½-inch solid wood exterior door has an R-value of about 2.2. An aluminum storm door can reduce heat losses by another 45 percent. This does not translate into a substantial dollar savings, however. Two average-size exterior doors lose about $34 worth of heat per year together. The addition of aluminum

DESTRATIFIER

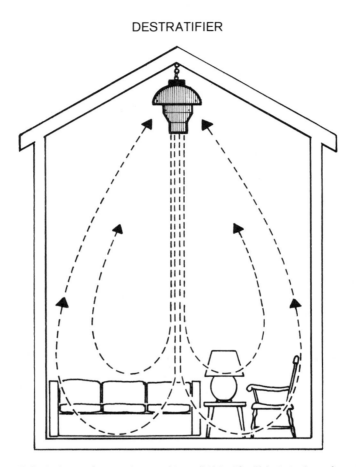

Figure 7-3. A destratifier can be used to pull "stratified" hot air down from the ceiling area and redistribute it near the floor where it can warm the people in the room.

storm doors reduces this loss to $27. You save $7 a year. This means that at $200 per door, it would take 50 years to get your money back.

It is possible to buy insulating or thermal doors that would provide a much higher overall R-value (i.e., R-3.2 for a regular door plus storm door versus R-10 for a thermal door). The cost of thermal doors is a bit higher, but they radically cut heat losses. You may also wish to consider building your own thermal door by sandwiching rigid insulation between panels of wood. The payback period on thermal doors is much shorter than on aluminum storm doors, but the actual savings to you are still less for this conservation measure than for the other steps on our list, so you should not give it a high priority.

The Impact of a Conservation Retrofit

Table 7-2 shows the economic impact of the conservation steps we have just listed. (Step 7 is not included since destratification is applicable only in homes with high ceilings.) The second column of the table shows how much homeowners in five locations around the nation might expect to pay for heat in electrically heated homes assuming these homes have not benefited from the conservation steps we have described. Columns three, four, and five show the cost of electric heat depending on how thoroughly conservation steps 1, 2, 3, 4, 5, and 8 are carried out. For example, a Long Island family that started out paying $5,685 a year for electric heat could cut their bills to $931 a year by insulating the ceiling, walls, and floors, by installing storm windows and insulating shades, by weather-stripping and caulking, and by adding storm doors and insulated doors.

Columns six and seven add the effect of setting back the thermostat (step 6). These columns show the cost of heat assuming that steps 1, 2, 3, 4, 5, and 8 have been fully carried out *and* the thermostat is set back 10°F at night (column six) or 15°F at night and 3°F during the day (column seven). The bottom line is that following the conservation steps we have described can significantly cut your heating bills. Thus, the Long Island family could lower its bill for electric heat all the way to $745, a total savings of $4,940 a year!

Other Options

Assuming you decide to carry out a comprehensive conservation retrofit of your house, and that you intend to stay with an electric system for most of your heat, there are supplemental heating systems you should consider. You could lower your heating bills by shifting part of your heat load to such a system. You might heat most of your rooms with your present electrical system, but you could heat a few rooms with a more energy-efficient supplemental system.

Various supplemental heating systems—having various strengths and weaknesses—are discussed in options 1 through 8, below. Based on the anticipated escalation in electric rates in many areas of the nation, however, and based on the fact that electricity is inappropriate as a heating source since it involves the conversion of high-grade energy into low-grade energy, I recommend the following: If at all possible, you should completely abandon electrical resistance heating and shift completely over to an alternative fuel, as I describe in options 9 and 10.

Option 1: The Heat Pump

Heat pumps intended to heat water were discussed in chapter 2. Heat pumps can also be used to heat the air in a home. The most common type of

Table 7-2

Electric Heating Costs for 1,500-Square-Foot Houses*

Location	Poor Conservation	Fair Conservation	Good Conservation	Excellent Conservation	Excellent Conservation plus 10°F Night Setback	Excellent Conservation plus 15°F Night Setback and 3°F Day Setback
Columbus, Ohio	$6,926	$2,548	$1,340	$1,134	$999	$907
Little Rock, Ark.	$3,805	$1,400	$737	$624	$549	$499
Long Island, N.Y.	$5,685	$2,091	$1,100	$931	$819	$745
Madison, Wis.	$9,305	$3,423	$1,800	$1,534	$1,341	$1,219
Salt Lake City, Utah	$7,163	$2,263	$1,386	$1,173	$1,032	$938

*Electric rate: 9¢ per kwh.

air-heating heat pump extracts heat from outdoor air and transfers it to the inside of your house. Although the outdoor air may feel cool to you, it usually contains sufficient heat energy for a heat pump to capture and use for interior space heating.

The advantage of the heat pump is its ability to provide two to three times more usable heat to your living space for a given number of kilowatt-hours than electric baseboard heating systems. Thus, compared to conventional electric heating systems, the heat pump can clearly cut your heating costs. Unfortunately, once late fall arrives, the outdoor temperatures in the colder regions of the United States drop to a point where heat pumps cannot operate efficiently and an electric heating coil kicks in. Thus, the heat pump really cuts your heating bill only in the spring and fall.

Many heat pumps are equipped for a reverse cycle that allows you to air condition as well as heat your home. If you wish to air condition and if you already have conventional electric heat, the heat pump can be a good buy: It will provide supplemental heat during most of the cold months, and it will provide air conditioning during the warm months.

A conservation retrofit can substantially reduce the annual consumption of electricity needed for space heating. A room-size heat pump could meet a substantial part of your remaining heating needs. Because such a heat pump allows zoned heating, if your family spends part of the day in the living room, the rest of the house can be kept much cooler and the heat pump can bring temperatures in the occupied area of the house up to a comfortable level. This approach can offer large savings during spring and fall.

There are also full-house, central heat pumps (analogous to central air conditioners). These units are not really practical in the Northeast because they are inoperative during the time of year when heat is needed most. They are also expensive—about $2,000, not including installation and duct work.

A room-size reverse-cycle heat pump/air conditioner with a high-energy efficiency (an EER of 9 or more), adequate for supplemental zone heating during the milder months, sells for $500 to $600 and is easily installed. An annual savings of 15 to 20 percent of your postconservation heating cost could be achieved. The payback period then would be about three years. In the milder areas of the country, savings could be substantially higher. Studies show a 25 to 50 percent average savings with central units. Two room-size models might achieve savings almost as high. In the warmest sections of the United States, you might meet your entire heat load with a heat pump.

For more information about this topic, see the section of part 2 titled "Air-to-Air Heat Pumps."

Option 2: A Portable Kerosene Heater

At $1.45 per gallon for kerosene and 8¢ a kwh for electricity used for heat, you get 31 percent more usable heat from each dollar spent on kerosene than

you do from each dollar spent on electricity, assuming the kerosene is burned at 60 percent efficiency. If you burn 150 gallons a year, it will cost you $218 to displace 3,561 kwh or about $285 worth of electricity, a savings of $67 a year. Thus, we could reduce our heating costs from approximately $787 to $720 a year in our example home.

If economics were the only consideration, portable kerosene heaters would seem to be justified. There are, however, certain problems associated with these heaters. One involves their potential fire hazard. Because they can be shifted around, they may be vulnerable to dangerous spills. Secondly, they tend to be messy and inconvenient to fill. They also adversely affect air quality in your house, because they vent their exhaust gases directly into the house. If you tighten your home so as to reduce the amount of infiltration, interior air pollution becomes even more of a problem. Although it is possible to operate kerosene heaters near a slightly open window, the heat losses due to infiltration are so great that much of what you gain is ultimately lost.

There are also maintenance costs and labor involved in getting the fuel and filling the heater frequently. Despite these drawbacks, if you are still interested, an investment of $200 to $250 could save you about 10 percent a year on electric heating costs.

Option 3: Electric Radiant Heaters for Task Heating

It is usually not necessary to heat an entire house to a uniform temperature. The use of zoning, discussed earlier, is an appropriate technique for conserving energy. A house can be heated to just 55°F with a primary heating system, while occupied rooms are heated to a comfortable 68°F to 70°F with a smaller secondary heating system.

Portable radiant heaters, including quartz heaters, take the zoning notion a step further. Instead of heating an entire room, you are warmed directly by radiant heat emitted by a smaller electric-resistance device. These heaters are ideal for close-range localized situations. In many ways, the effect is similar to sitting around a campfire.

If you sit or work in a particular spot for extended periods, you may wish to consider radiant heaters. They draw as little as 600 to 800 watts per hour, but they do a pretty good job of directing heat toward you with aluminized reflectors.

Energy savings would of course vary, but a hypothetical situation might be as follows: A household is inhabited by two adults and two teenagers. At 8:00 P.M., the thermostat for the primary heating system is automatically set to 55°F, since all the residents are gone by 8:30. The teenagers return home at 3:30 P.M., the adults at 5:30. The automatic thermostat could bring the house temperature up to 70°F by 3:30 for the returning teenagers. However, the kids normally do homework or watch TV in their rooms from 3:30 to 5:00. The

thermostat is therefore set for only 62°F at 3:30 and for 70°F at 5:00. During the hour-and-a-half interim, the teenagers each use a radiant heater to warm them in their rooms. To heat the entire house to 70°F for this time period every day would be far more expensive than keeping the house temperature down and using radiant heaters.

The overall impact on your heating bill is difficult to determine because individual use of these devices will vary substantially. Keep in mind, however, that thermostat setbacks provide a large savings in overall heating costs, and if you can extend the duration of these low-temperature periods, you could save a great deal of money. Radiant heaters sell for $30 to over $200 and they come in many varieties. If you are really interested in one or several of these devices, do some comparative evaluations before you buy. Because they are portable, radiant heaters must be handled with care. They are a fire hazard when operated near flammable materials, and young children could suffer burns or electrical shocks from them. Be sure that family members are responsible enough to use these devices.

Option 4: Minifurnaces

Minifurnaces burn kerosene, gas, or oil, and they come in three basic types—direct-venting, chimney-venting, and nonventing. All are meant to be located in your living space.

Direct-venting models are extremely efficient because they draw their oxygen for combustion directly from outside air rather than from the interior of your house. Instead of venting through a chimney, they vent their exhaust gases through a round tube that extends from the heater through the house's wall. Air for combustion is drawn into the furnace through a larger tube that encircles the first tube. This way high efficiency (90 percent) is achieved without causing indoor pollution. The output of direct-venting minifurnaces ranges from a capacity of about 10,000 Btu per hour to over 30,000 Btu per hour. The units frequently come with electric blowers to disperse heat over a larger area and with wall thermostats and clock timers.

Another variety of minifurnace is meant to be used with a masonry or metal chimney. These are less expensive than direct-venting models but are also less efficient since they—like most conventional heating systems—draw on warm inside air for combustion and vent some of this warmth to the outside environment. They have a capacity of up to 50,000 Btu per hour, but since they run at 60 to 65 percent efficiency, much of this heat energy is lost. Still, such a minifurnace will warm you for far less than an electric heating system.

The third type, nonventing minifurnaces, will deplete some of the oxygen supply in your house. They are similar to portable kerosene heaters in that they vent their exhaust gases into the house, thus creating indoor air pollution. I would not recommend unvented heaters.

DIRECT-VENTING MINIFURNACE

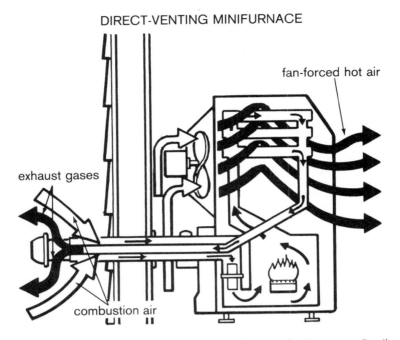

Figure 7-4. Direct-venting minifurnaces send their combustion gases directly outdoors, so they provide heat without polluting indoor air.

Direct-venting heaters, because they are so efficient, should certainly be considered as a supplemental heat source. They are more expensive than chimney-venting models but may be worth it over the long run. There are several high-quality kerosene-fired units available with capacities of 19,000 Btu to 32,000 Btu. Note that kerosene burned in less efficient equipment is not all that economical, but the high efficiency of direct-venting minifurnaces makes kerosene a viable substitute fuel for electricity. The units can be linked to a large remote fuel tank so that long-term use without constant refilling is possible. They cost a bit less than $1,000 and are not overly difficult to install.

Direct-venting gas-fueled minifurnaces are much less expensive than kerosene models—about $250 to $450. They do require an outside gas line hookup, but this is not a major problem unless natural gas is unavailable in your area. Remember, LP gas is far more expensive than natural gas.

Chimney-vented minifurnaces are even cheaper to buy—about $100 to $350. Although less efficient, they still offer major savings due to the change from electricity to oil or gas. Consider also that unlike the heat pump, which does not provide supplemental heat during the coldest months, all minifurnaces can be operated throughout the heating year.

Let's consider briefly the potential impact of a minifurnace on your heating bill. Sears sells a chimney-vented oil minifurnace having a 50,000-Btu capacity for about $170. Assuming it operates at 60 percent efficiency, we get 30,000 Btu of heat when it is running full blast. January is the coldest month of the year. Our example home, having been brought up to just a "fair" conservation level (see table 7-1), consumes almost 20 percent of the year's heat load during January—about 5,748 kwh worth of heat energy at a cost of $460. Each day in January, our average requirement for heat is 185 kwh or 7.7 kwh per hour. Thus, the amount of heat needed for an average hour in January is about 26,300 Btu. (There are 3,412 Btu in 1 kwh.) This means that during the coldest month, the minifurnace has more than enough heating ability (30,000 Btu) to meet the average requirement of the house.

If our example home were at an excellent level of conservation (table 7-1), we would use $193 worth of electricity in January. This is only 78 kwh per day or 11,089 Btu per hour. A minifurnace with far less capacity—about 15,000 to 20,000 Btu—would just about meet the average hourly heat requirement, assuming an efficiency of 60 percent (i.e., about 10,000 Btu per hour of usable heat energy).

Does this mean a minifurnace installed in your living space can completely eliminate the need to use electric heat? The answer is theoretically yes, but there are three qualifications to be made.

First, you should do a conservation retrofit to make your home energy efficient. Before spending money for fuel to heat the home, you should make sure that the home will retain the heat.

Second, you must be able to distribute the heat throughout the house. Other heating systems use either ducts or separate radiators to direct heat to all areas of a house. A 1,500-square-foot single-story home sprawls over a large area, and heat distribution without some auxiliary system to move heated air around would be problematic. Several low-wattage fans could help but probably would not be sufficient to get warm air to the extremities of your interior space. Several smaller furnaces with lower capacities placed in separate locations may do the trick, but you would have to contend with the hassle of several remote tanks or extensive fuel lines.

Third, a house's heating needs increase during periods of extreme cold— they may then exceed a minifurnace's capacity. The home's conventional electric system would have to be used in conjunction with the minifurnace to meet the heat load.

Despite these qualifications, we can say that a minifurnace, when used after a conservation retrofit, could easily cut electric demand by 65 percent, sometimes more. The remaining electrical usage would be for heating outer rooms and to supplement the minifurnace when temperatures are very low. If our example house were brought to an excellent level of conservation, and if 65 percent of the heat load were met with with oil (at $1.20 per gallon) burned at 60 percent efficiency while 35 percent of the heat load were met with electricity, our annual heating cost would be about $590, a savings of about $197

a year. Investing $200 in a chimney-vented minifurnace would involve a payback period of just one year. An $850 kerosene direct-venting heater would allow us to meet our annual demand for $542, assuming it met 65 percent of the load, the equipment operated at 88 percent efficiency, and the cost of fuel was $1.45 per gallon. Based on an annual savings of $245, the payback period would be about 2½ years, and no chimney would be needed.

For more information about minifurnaces, see the section of part 2 titled "Energy-Efficient Furnaces."

Option 5: Direct Passive-Solar Gain

Most of the scientific discussion about solar energy concerns complex, expensive, and often unreliable "active" systems. Active-solar systems rely on electrical and mechanical components to pump, monitor, and control energy flowing between a home and its solar-collector panels.

"Passive" solar heating, by comparison, tends to be more reliable and cost-effective and uses no mechanical or electrical components. The goal is to simply capture free energy from the sun. "Direct-gain" passive-solar systems do this by using windows on the south wall of a house as heating devices. South-facing double-glazed windows can provide substantial amounts of solar heat. Generally speaking, the windows should be fixed and well caulked. They should also be equipped with insulating shutters or shades to prevent heat losses when the sun isn't shining.

If you were to install 100 square feet of windows on the south side of our example house (standard patio door replacements are ideal and also inexpensive), you would receive the equivalent of about 36 kwh per year for each square foot of glass, assuming the sun is not blocked by trees or other buildings as it passes overhead in winter. The windows would intercept a total of 3,690 kwh worth of heat energy, which is $295 worth of electricity at 8¢ per kwh.

Table 7-3 shows the economic impact of installing a passive-solar direct-gain system on the south side of a 1,500-square-foot house. The system is simple, using 100 square feet of double-pane fixed-glass windows and some type of movable window insulation to reduce heat loss during sunless periods.

Column two of table 7-3 provides savings data for a typical, moderately insulated American house. Such a home would fall about halfway between the "fair" and "good" insulation status categories described in table 7-1. In column four, we see the impact of a passive-solar system on a house with "excellent" insulation.

Table 7-3 assumes that the price of adding 100 square feet of fixed double glazing and accompanying movable insulation is about $1,500. This figure includes $750 for glazing and moldings, $400 for movable insulation, and $350 for installation. If you were to purchase patio door replacements (the cheapest way to buy large double-pane panels), do the installation yourself, purchase inexpensive shutters or curtains or make them yourself, the total cost of your

		House with Moderate Insulation		House with Excellent Insulation	
Location	Electric Rate Per kwh* (¢)	Annual Reduction in Heating Cost ($)	Payback (yr.)	Annual Reduction in Heating Cost ($)	Payback (yr.)
Albuquerque, N. Mex.	7	349	4.3	326	4.6
Atlanta, Ga.	8	281	5.3	254	5.9
Bismarck, N. Dak.	7	392	3.8	333	4.5
Boston, Mass.	10	401	3.7	356	4.2
Burlington, Vt.	6	275	5.5	229	6.6
Cleveland, Ohio	10	402	3.7	400	3.7
Corvallis, Oreg.	4	180	8.3	161	8.3
East Lansing, Mich.	6	262	5.7	232	6.5
Fort Worth, Tex.	8	274	5.5	229	6.6
Fresno, Calif.	7	248	6.0	221	6.8
Indianapolis, Ind.	6	244	6.1	216	6.9
Las Vegas, Nev.	8	343	4.4	311	4.8
Lexington, Ky.	6	200	7.5	185	8.1
Little Rock, Ark.	8	265	5.7	238	6.3
Los Angeles, Calif.	8	385	3.9	317	4.7
Madison, Wis.	8	398	3.8	352	4.3
Manhattan, Kans.	8	328	4.6	292	5.1
New York, N.Y.	15	549	2.7	498	3.0
Oklahoma City, Ok.	6	239	6.3	212	7.1
Phoenix, Ariz.	8	296	5.1	250	6.0
Raleigh, N.C.	7	254	5.9	225	6.7
Reno, Nev.	9	496	3.0	457	3.3
Salt Lake City, Utah	8	402	3.7	389	3.9
San Antonio, Tex.	8	230	6.5	202	7.4
Shreveport, La.	6	176	8.5	159	9.4
Spokane, Wash.	3	150	10.0	131	11.5
State College, Pa.	5	200	7.5	173	8.7
Tampa, Fla.	6	133	11.3	94	16.0
Washington, D.C.	6	201	7.5	185	8.1

Table 7-3

Impact of Adding 100-Square-Foot Passive-Solar Direct-Gain System to 1,500-Square-Foot House

*Based on average local electric rates, minus a 1¢ discount such as many electric companies offer to users of electric heat.

direct-gain system would be less. If more expensive materials were used or if you were to borrow the money for the retrofit at high interest rates, it would cost you more.

Table 7-3 should help you decide whether a passive-solar retrofit of your home is worth doing. Locations with the fastest payback times are the best bets for carrying out a solar retrofit. Even if your particular area is not given, certain universal factors emerge in the table. First, if you live in the coldest regions of the country and if your local electric rate is 8¢ or more per kwh, adding glass to the south side of your home would bring you substantial savings each year and your payback period would be around four years. In very cold areas, the retrofit is worth doing even if electric rates are only 6¢ or 7¢ per kwh.

Secondly, if your electric rate is moderate or above (i.e., 7¢ or more), and if you live in a high-sunshine area—the lower South, the Southwest, southern California, or the southern section of the Rocky Mountain West—a solar retrofit is worth doing if you can save $200 or more per year (i.e., if your payback period is less than seven years). If your payback period is longer than seven years, as for example in southern Florida, the investment is only marginally justified. Thus, in the cities of Albuquerque, Atlanta, Fort Worth, Las Vegas, Los Angeles, Reno, and Salt Lake City, where the annual percentage of sunshine is very favorable, savings on electric heating costs should be returned in four to six years. Keep in mind, however, that extensive south glazing in warmer areas of the United States can mean substantial solar heat when you do not want it as well as when you do. Without proper overhangs, awnings, and other sun-blocking devices on or around your home, you may defeat the purpose of saving energy by driving up summer cooling bills.

The regions where the addition of solar glazing is least justified are those where electric rates are less than 7¢ and where heat loads are only moderate or the amount of winter sunshine is low. The Northwest tends to have very low electric rates, moderate heating needs, and poor sunshine. A direct-gain system is generally not economically justified in that region.

In our well-insulated example home in Columbia, Missouri, the annual electric heating bill, after we add 100 square feet of glass to the south side of the dwelling, drops from $787 to $496, a savings of $291 a year. Based on an initial investment of $1,500, our outlay would be repaid in about five years. This is a remarkably low heating cost, considering that we are still using a baseboard resistance system as our primary heating source.

For more information about direct-gain solar heating, see the section of part 2 titled "Passive-Solar Space Heating."

Option 6: The Passive-Solar Attached Greenhouse

A wide selection of prefabricated solar greenhouses is now available to the homeowner. A solar greenhouse intercepts solar energy in substantial quantities.

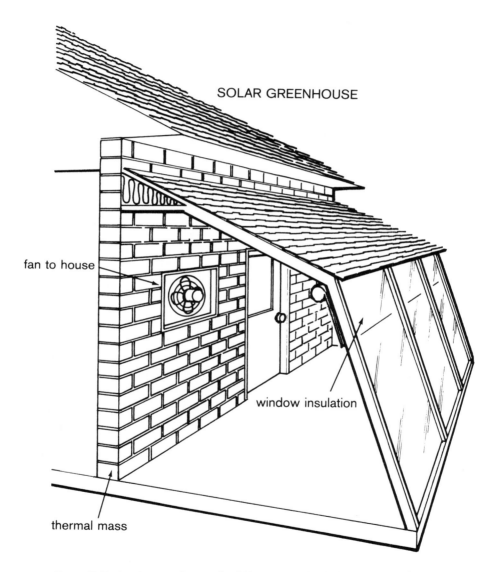

Figure 7-5. A solar greenhouse should be equipped with thermal mass (such as masonry or water-filled containers) to store heat and window insulation to prevent heat from escaping through the glazing.

It is not uncommon for greenhouses to reach 125°F on a winter day. The greenhouse sends its excess heat into the home's living space. A low-wattage fan can be used to facilitate the movement of this solar-heated air into your house.

Solar greenhouses must either be sealed off from the house at night to prevent heat loss or they must be fitted with insulating shades or shutters. Good greenhouses are double glazed, tightly sealed, and well insulated.

How much solar energy you can capture depends on where you live, the size of the greenhouse, the angle of the glazing (i.e., whether it is vertical or sloped), the quality of construction and materials, and the greenhouse's orientation toward south. How efficiently the greenhouse stores and/or transfers heat will also influence how much heat you actually get. A 14-foot-long by 11½-foot-wide greenhouse attached to a well-insulated house can meet as much as 30 to 40 percent of the home's annual heat load in the northern United States and 60 to 80 percent in some parts of the South.

In Columbia, Missouri, the annual cost to electrically heat a well-insulated, tight home equipped with a thermostat control device should drop from $787 to $475 after we add a solar greenhouse on the south side. This is a savings of over $300 a year based on a contribution by the greenhouse of 40 percent of the home's heat load.

Assuming a cost of $5,000 for the greenhouse (costs can vary from $2,000 for the cheapest models with no extras up to $12,000 for fully equipped units) the payback period is 17 years. Thus, it would be difficult to justify investing in a greenhouse unless you wish to also use it for additional living space or as a place to grow plants in the winter. In cold but sunny climates where electricity rates are high, a greenhouse can be justifiable, especially if it is purchased at the lower end of the price range.

Besides providing additional living and plant-growing space, attached greenhouses have another advantage over direct-gain systems. Because a solar greenhouse raises the temperature of the air outside the wall to which it is attached, heat losses through that wall and from the building itself are reduced. This actually lowers the annual heat load of the house (i.e., you will need less fuel to stay warm). Direct-gain passive-solar systems, on the other hand, require extensive glazing to be installed in the home's south wall; thus, a surface with a low R-value is added to the dwelling. Heat losses through a double-pane glass wall can significantly reduce the net energy impact of a direct-gain system, especially in a region with a cold climate and a great deal of intermittent cloudiness. As a general rule, it can be stated that an attached solar greenhouse, when properly designed, constructed, and used in conjunction with a well-insulated south wall, is more appropriate in the northern United States than is a direct-gain system. The cost of most greenhouses is high, however.

For more information about solar greenhouses, see the section of part 2 titled "Passive-Solar Space Heating."

Option 7: Solar TAP Collectors or Window Heaters

Easy-to-install solar collectors are now available. They are designed to be mounted on your exterior south wall or connected to the bottom of a double-hung window. Units vary in size but typically have an area of 16 to 24 square feet. Their seasonal heat output can vary over a wide range, depending on the design and efficiency of the collector and the local solar and climatic conditions. The systems provide anywhere from 30,000 to 120,000 Btu per square foot of collector area each season. In Columbia, Missouri, a good system could provide about 55,000 to 75,000 Btu per square foot of collector over the winter months, which means one collector could provide the equivalent of about 325 to 440 kwh, based on a 25 percent efficiency. Four of these solar collectors could contribute 20 percent of the annual heat load at our example house, assuming the house is well insulated.

A typical "TAP" collector (otherwise known as a thermosiphoning air panel) is mounted on the south wall of a house. It is an insulated box containing an absorber plate and single or double glazing on the side that faces south. The absorber plate, which is inside the box behind the glass, is a sheet of metal that becomes hot as the sun shines on it, thus "absorbing" solar heat. The plate is painted a nonreflective black to maximize the amount of heat it can absorb. Air from inside the house enters the box through a vent cut through the wall at the bottom of the collector. When the air comes into contact with the absorber plate, it is warmed and rises up and out of the box through another vent, thus carrying heat into the house. A cycle is set up: Cool air enters the box, is heated, rises, and makes way for more cool air, which is drawn into the lower vent of the TAP. This process is called thermosiphoning—hence the name, thermosiphoning air panels.

Air that enters a TAP at 65°F exits at 90°F or 95°F. Some units use only the natural tendency of rising hot air to maintain the flow of air into and out of the box; others use an electric fan to speed up the process. A fan boosts the overall efficiency of a TAP by increasing the volume of air that passes over the absorber plate. There is less time for the air to become hot, so the air's exit temperature is perhaps 10°F less—about 80°F or so. Nevertheless, a much larger volume of warmed air is sent into the living space.

Window box heaters work on the same principle as TAPs except that they are designed to be incorporated into a normal window on the south wall of a home. Thus, you do not need to cut vents in your wall. The collector is a box with a fan in it, similar in shape to a room air conditioner although smaller in size. The window is opened, then the collector is inserted into the window opening. The window is then closed onto the top of the unit and caulk is used to seal any seams or openings. On sunny days, the unit will provide ongoing free heat energy.

TAPs and box heaters run about $300 to $700 each. You can also buy plans and construct them yourself for less than $200. The average 20-square-

SMALL SOLAR HEATERS

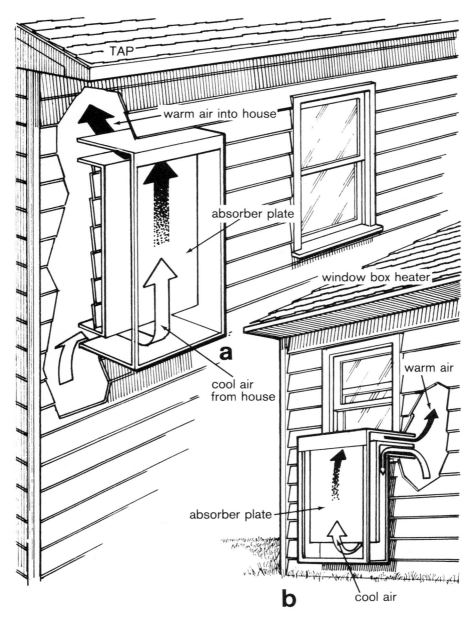

TAP

warm air into house

absorber plate

window box heater

a

cool air
from house

warm air

absorber plate

b   cool air

Figure 7-6. A ''TAP'' collector (a) can be attached to a house's wall. Cool air
from the house enters the collector, is heated by the sun, and returns to the
house. A window box heater (b) is similar, except that it is hung from a window
instead of attached to a wall.

foot unit should save you about $35 a year on electric heat. Solar collectors are now labeled with Btu ratings. If you divide the Btu rating for a specific model by its cost, you will be able to determine how much heat you can hope get per dollar invested. There are over 500 collectors on the market. Compare before you buy.

Two final notes concerning all forms of solar heating systems. Don't be misled by the total output of heat energy you can *potentially* capture from the sun each year. Since the sun only shines in the day but your heat load is much greater during the sunless nighttime hours, a solar heating system may produce more heat than you need during the day while leaving you without sufficient heat at night. The way to combat this is to incorporate "thermal mass" into your home—masonry or other heavy material (e.g., water-filled containers) that will soak up the heat during the day then release it gradually during the night. Without thermal mass, you may waste your solar heat through daytime overheating. This means that solar space heating systems should not be sized to provide more than comfortable daytime temperatures unless you have a means of storing up those extra Btu.

As is the case for conventional heating equipment, a method for distributing heat throughout the house is necessary. Otherwise, a solar heating system may provide heat to only a small area of the house. Low-wattage fans are ideal for this purpose. For example, an exhaust fan can efficiently move hot air from a solar greenhouse into nearby rooms, and additional small fans can be used to direct warmth to other rooms. Low-speed fans are preferable, since high-speed models will generate a cooling draft.

For more information about TAPs and window heaters, see the section of part 2 titled "Passive-Solar Space Heating."

Option 8: Install an Airtight Wood Stove

Wood could make a substantial contribution to the nation's heat needs if it were burned in efficient stoves and used to heat tight houses. It is already our fourth largest fuel after oil, natural gas, and electricity. A well-designed and well-built 1,500-square-foot solar-heated house may require only about one cord of wood per year to provide supplementary heat. At this level of consumption, there is plenty of wood to go around.

A cord is a measure of volume—128 cubic feet. How much heat you get from a cord will vary according to the density of the wood, how much moisture it contains, how much air space there is between the stacked pieces, and the efficiency of the stove in which it is burned. An average cord of wood is 33 percent air, and its potential heat energy ranges from 13 to 39 million Btu, with an average of about 20 million Btu. Quality wood stoves are airtight and about 60 percent efficient, so you can expect to get about 12 million usable Btu from a cord of fuel wood burned in such a stove—the equivalent of about 3,500 kwh or $280 worth of electric heat at 8¢ per kwh. Since the average price per cord of wood in the U.S. is between $80 and $160, wood is clearly much cheaper than electricity as a heat source.

A wood stove with the appropriate Btu rating for the size of your house is capable of meeting your house's total heat load. However, as with the minifurnaces and solar heating systems discussed above, the distribution of heat to extremities of the house is a problem. Periods of extremely low temperatures may also require a back-up supplemental heat source. Most families use wood stoves to meet a large part, but not all, of their heating requirement.

If, at our example home, we burn two cords of wood at 60 percent efficiency and pay $120 for each cord, we get the equivalent of 7,000 kwh of heat. Since the home's total heat load after a conservation retrofit is 9,838 kwh, we need another 2,838 kwh worth to heat, which we could get from the home's electric heating system. Thus, we would have an annual cost of $240 for wood plus $227 for electricity, or $467 total.

If we wanted to heat our example house almost entirely with wood, we would need a wood stove of about 15,000- to 20,000-Btu capacity. This gives us about 9,000 to 12,000 usable Btu, enough to meet the house's heat load on all but the coldest days, when the house's electric system would provide supplemental heat. A still larger stove could generate even more heat and meet extreme day loads, but it would be oversized for most days of the winter.

A high-quality European or American airtight wood stove of 10,000-Btu output, with a metal chimney and installation, should cost $1,200 at the most. If you save $250 to $300 a year on heating costs, you would recover your investment in four years. Unlike other heating systems, there is a good deal of "trading up" among wood stove owners (i.e., moving to bigger stoves or to coal). Thus there is a sizable market in used wood stoves. Many excellent stoves can be bought at reasonable prices. You should consider this when you look around.

Several innovative stove designs have recently emerged and it is now possible to achieve heating efficiencies of 70 percent or more. Catalytic stoves, for example, work by reducing the temperature at which woodsmoke burns. The smoke thus tends to burn, releasing more heat, before going up the chimney.

For more information on this topic, see the section of part 2 titled "Wood Stoves."

Option 9: Install a Coal Furnace or Stove

Unlike wood, coal is not renewable. There is, however, plenty of it around. If supply alone determined the price of coal, it should go down rather than up as it has over the past few years. But since coal fields are controlled by the same corporations that control oil and nuclear fuels, the price seems bound to rise. Nonetheless, coal at $140 per ton offers about the same number of usable

CATALYTIC WOOD STOVE

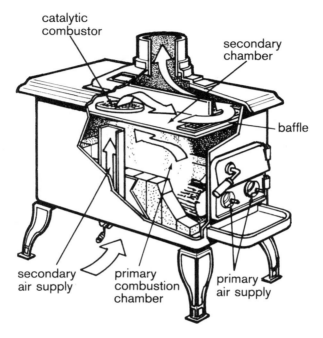

catalytic
combustor

secondary
chamber

baffle

secondary
air supply

primary
combustion
chamber

primary
air supply

Figure 7-7. A catalytic wood stove extracts heat from burning wood, and then extracts additional heat from the smoke rising from the wood fire.

Btu per dollar as wood does. It is an economic bargain, although far from an ecological one.

One ton of coal burned at 60 percent efficiency gives as much heat as 4,748 kwh of electricity. The former costs $140, the latter $380. Thus, conversion to coal as the only heat source at our example house could bring heat costs down to $290 a year.

Despite its cheapness, coal does have drawbacks. It is relatively dirty, polluting outdoor and sometimes indoor air. The ashes are difficult to get rid of, and heavy buckets full of ash accumulate quickly. Moreover, coal fires are hard to start, and, in the milder periods of the year, coal stoves usually put out more heat than you need or want.

A number of people have converted to coal as a supplemental heat source while retaining their existing primary systems. Assuming a reduction in the cost of electric heating to $787 per year at our example house, 1½ tons of coal at $140 per ton burned in a small room-size coal stove would mean a combined electric/coal fuel bill of $427 per year.

Option 10: Convert to an Energy-Efficient Oil or Gas Furnace

As I stated above, electricity is an inappropriate fuel for space heating. The ideal solution to your heating problems would be to stop using electricity altogether. You could switch to coal (option 9), or to some combination of wood stoves, solar heating, minifurnaces, etc. Or you could switch to a gas- or oil-fired furnace.

Based on the current costs of natural gas and oil, a high-efficiency furnace is probably your soundest energy-saving choice after the completion of a full conservation retrofit of your home. Years from now, with nonrenewable fuels being subject to unpredictable price fluctuations, I suspect that solar and wood systems will enjoy greater prominence than they do today. But for now, gas and oil seem the most practical.

How much heat a furnace delivers, as compared to how much heat is potentially available in the heating fuel, is an important issue. The efficiency of your furnace reflects how well it extracts heat from your heating fuel for use in warming your living space. A conventional older oil furnace before a tune-up will be about 50 percent to 60 percent efficient. A tuned oil burner should be 60 percent to 70 percent efficient. Oil furnaces with conservation devices added may reach 70 to 75 percent, and new superefficient models are about 85 percent efficient. The efficiency ratings of new equipment are shown in a U.S. Department of Energy rating known as its Annual Fuel Utilization Efficiency (AFUE).

Let's consider briefly what these figures can mean to our heating bill. Our example home needs $787 a year worth of power for electric heating. This much electricity contains around 33.6 million Btu. The identical house with a ten-year-old untuned oil furnace would require $580 worth of #2 fuel oil costing $1.20 per gallon. Notice that even with this incredibly inefficient heating system, we still would pay about $200 a year less to heat our example home with oil than with electricity in a resistance system.

If we installed the most efficient oil furnace available (AFUE 87.8) in our home, we would meet the annual heating requirement for just $330 a year, based on the $1.20 per gallon for fuel oil figure. Thus, we would save over $450 annually.

If we installed the most efficient natural gas furnace available (AFUE 97) and purchased natural gas at $7 per 1,000 cubic feet, our annual heating cost would be $231, for an annual savings of more than $550!

The inappropriateness of electricity as a fuel for heating is reflected in the savings achievable with oil and gas. When we introduce a gas furnace with an efficiency approximately equal to that of an electric baseboard heating system, it becomes clear that electricity is a luxury fuel costing almost three and a half times as much as natural gas for delivered Btu.

For more information about oil and gas furnaces, see the section of part 2 titled "Energy-Efficient Furnaces."

Recommendations

1 In the colder northern regions: Do a complete conservation retrofit. Purchase an efficient oil or gas furnace (option 10) and install an automatic thermostat timer. The total cost should be about $3,000 to $4,000 and could save you approximately 70 to 80 percent each year compared to present electric heating costs if your home is moderately insulated at the present time and if you have an older-model furnace. If your annual heating bill has been around $2,000, it could very well drop to $350 to $450. The payback period would then be less than three years. If your home is already well insulated, you can still expect to save $400 to $500 a year with a payback period of three to four years on a superefficient oil or gas furnace. This recommendation is probably the most desirable one for most Americans, since it allows the homeowner to continue heating conveniently with no additional effort required by the household members.

2 In the colder regions (except the Northwest), in high-sunshine areas, and where electric rates are now or will soon be 8¢ per kwh or more: Do a complete conservation retrofit. Install a direct-gain passive-solar system or an attached greenhouse with insulating shutters or curtains. In the southern sections of the United States and in temperate zones, you may want to continue using your electric baseboard system as your backup heat source or shift to a heat pump for your backup system. In this case, your overall bill should be reduced by between 60 and 80 percent if your home is presently insulated to a moderate level.

In New England, the mid-Atlantic states, and forested regions, you might wish to convert from your electric system to an efficient wood stove, or you might want to consider an oil, gas, or kerosene direct-venting minifurnace. Any of these systems can be combined with passive solar for a low-cost heating package. If you heat with solar and wood, you will be meeting your heat load solely with renewable energy sources at an annual cost of $500 a year or less in the coldest regions. An investment of $6,000 for conservation, direct-gain solar, and a wood stove or minifurnace would be returned in savings in about four years in the North, a bit longer in milder sections of the country. This option requires some degree of activity to operate insulating shutters and may require fuel handling, so it is not suited to everyone.

3 All regions: Do a complete conservation retrofit. Convert to an efficient minifurnace that uses a less expensive fuel than electricity and install several small fans to help promote the even distribution of heat throughout the house. Your baseboard system can always be used as a backup when extreme tem-

peratures prevail. This option is cheaper than the two above and can still reduce heating costs by 60 to 70 percent.

The Bottom Line

Let's consider a final illustration that I hope will convince you of the merits of taking direct action toward resolving your current electric heat situation.

To carry out a $5,000 energy-reduction program, let's assume you will have to borrow the money at 14.75 percent and repay it over three years at $175 per month or $2,100 each year, including interest. The impact of your action will be a drop in your annual heating cost to, say, $500. Let's also assume that your house presently has just a fair amount of insulation. Thus, you now suffer a yearly heating bill of $2,265, based on a rate of 8¢ per kwh for electric heat.

Table 7-4 outlines the savings you will achieve. I assume that for your $5,000 investment, you buy a complete conservation retrofit and that you convert to one

Table 7-4

Cost to Heat Moderately Insulated House

Year	Electric Rate per kwh* (c)	Cost to Heat without Retrofit ($)	Loan Repayment ($)	Cost to Heat with Retrofit† ($)	Total Annual Outlay ($)	Annual Savings ($)
1	8.0	2,265	2,100	500	2,600	− 335
2	8.4	2,378	2,100	525	2,625	− 247
3	8.8	2,492	2,100	550	2,650	− 158
4	9.2	2,605	0	575	575	2,030
5	9.6	2,718	0	600	600	2,118
6	10.0	2,831	0	625	625	2,206
7	10.4	2,945	0	650	650	2,295
8	10.8	3,058	0	675	675	2,383
9	11.2	3,171	0	700	700	2,471
10	11.6	3,284	0	725	725	2,559
Average	9.8	2,775	—	613	1,242	1,532

Total Savings = 15,322

*Assumes .4¢ annual rate increase (i.e., 45 percent over the decade).
†Assumes $25 annual fuel cost increase (i.e., 45 percent over decade) for oil, gas, or wood.

of several less expensive heating technologies (gas-, oil-, or wood-burning systems with high operating efficiencies).

The net savings over a ten-year period due to the conservation retrofit/energy system upgrade is $15,322. If rate increases are for some reason less than .4¢ a year, the net savings would be reduced, but the total savings should still be in the range of $12,000 to $14,000.

My advice to you is, therefore: Borrow some money, if necessary; retrofit your house; and change from electric to another heat source. The economics are overwhelmingly against continuing with electric heat.

Chapter 8
Other Appliances

Electricity is ideal for running motors. When used to generate heat, it is an economic disaster. Any appliance containing a high-resistance coil is bound to separate you from your money at a furious clip. Your biggest potential electricity savings related to appliances, therefore, will come from eliminating or substantially reducing the use of heating coils in home appliances. Wise use of appliances should net you a 40 to 50 percent savings over current costs. Let's consider the possibilities.

Clothes Dryer

A typical electric dryer costs about 46¢ an hour to operate at 9¢ per kwh. The majority of this cost is related to heat generation; only a small part is related to the energy used to run the motor. The average household uses a clothes dryer 210 hours a year at a cost of $97. You can save on this expense by following one or several of the suggestions given below.

☐ Convert to a solar dryer (i.e., a clothesline). It may sound primitive, but during much of the year hanging clothes outside to dry can be an enjoyable experience that provides a 100 percent energy savings on every wash load you don't dry electrically. If you hang your clothes outdoors during the warm months, you could save about two-thirds or roughly $65 a year.

2 Install an indoor drying rack for use during inclement weather. It can be set up in a variety of indoor locations and will also help to boost indoor moisture levels in winter. In this way you save on the operation of the dryer and of a home humidifier as well.

3 Use your automatic clothes dryer wisely. If you fill your dryer, you will get more dry clothes per dollar spent than if you under or overfill your machine. Keep your exhaust outlet and lint screen clean, thereby boosting your dryer's efficiency. If your machine has an automatic dry cycle, set it so that it does not run longer than necessary. Thus, a setting of 40 minutes may be adequate for lightweight clothes, and an hour should be enough for heavier items. You can always start the dryer again if clothes are not completely dry; this is a more economical approach than allowing your dryer to run until you remember to turn it off. Dry loads consecutively to take advantage of built-up heat. Separate loads into heavy and light items. These practices could save you about 10 to 15 percent a year.

4 Vent the exhaust from your electric clothes dryer indoors in winter. Almost 50 percent of the heat your dryer produces is dumped uselessly outdoors. Why not direct the warmth into your home where it can be put to good use? This will not reduce your drying bill, but it could lower your heating bill. The money to pay both bills all comes out of the same pocket. A simple-to-install device called a "dryer heat reclaimer," costing $5 to $30, will make winter interior venting and summer exterior venting easy to do.

5 Convert to a gas dryer when your electric model wears out. If you pay above 9¢ per kwh, you should save about two-thirds by converting to a new energy-efficient gas dryer with electronic ignition. A $320 model should pay for itself in about four years.

Dishwasher

Because motors are relatively efficient devices, the high energy consumption of dishwashers is related to their use of hot water and the electric drying element they contain. A typical dishwasher uses approximately 2,100 kwh a year.

1 About 1,740 kwh (83 percent) of a typical dishwasher's electric consumption goes for heating water. Recommendations on reducing hot water use were given earlier. You should review them. Keep in mind that if you turn down the thermostat on your water heater, you may have to boost the water temperature again at your automatic dishwashing machine. Some models have a built-in booster, or you may have to purchase a small point-of-use water heater to raise water temperature to a satisfactory level. These heaters cost about $150 but are worth it since they allow you to reduce overall water temperatures at the tank.

2 About 360 kwh per year are consumed by a typical dishwasher for purposes other than water heating; up to half of these relate to the drying cycles.

Newer dishwashers should have a switch that allows you to skip the drying cycle. Use this switch if you have it. Let your dishes air-dry instead. If your machine does not have this switch, disconnecting the heating element is relatively easy. This may save you $10 to $13 a year.

3 If your machine has an optional short-washing cycle, you'll be able to reduce the amount of hot water used by about one-third. This could save you 575 kwh a year or about $52, if you heat water electrically.

4 Many dishes and glasses don't really need more than a light rinse, so why run them through the dishwasher? The less detergent you use, the better for your body. Read the label on the box: It indicates that the ingredients are caustic. Every time you wash, a residue remains that winds up inside you.

5 Do only full loads and you'll save water. If your machine can use 120°F water, you'll also save. And doing dishes by hand could save you 50 percent or more per load.

Clothes Washer

A typical automatic clothes washer motor uses only 100 kwh a year. The water it uses, if electrically heated, requires another 2,400 kwh a year. Thus, at $9 a year, your clothes washer is a bargain to operate, and any large savings would have to come from reducing the quantity of hot water the machine requires and/or lowering the cost of heating that water. As noted earlier, the conversion to a gas hot water system substantially lowers water-heating costs.

As far as your machine is concerned, you can cut costs by taking these steps:

1 Do full loads, not more, not less. This gives you optimal washing for your dollar.

2 Convert to colder water levels for both washing and rinsing. A warm wash and cold rinse are adequate for most clothes washing.

3 When you buy your next machine, make sure you have a "suds saver" or "water saving" option. This allows you to reuse your wash water for a second or even a third consecutive load. You should also look for a machine with a minibasket option for doing small loads.

For more information about clothes washers, dryers, and dishwashers, see the appropriate section of part 2.

Television

Most of the newer television sets are "instant-on" models. This means you get a picture the second you turn your set on. Very convenient, but a large waste of your money. At 9¢ per kwh, it costs you about $15 to $20 a year to avoid warm-up time. The overall cost of running a typical set is about $35 to $40 a year. Doing without the instant-on feature could save you 40 percent of your annual outlay for TV power. If you have a second set that is used only half as much as your main set, you could save 55 percent on the cost of running it by eliminating its instant-on feature.

To eliminate instant-on, you have to cut off power to your television set when it is not in use. This can be done most simply by pulling the plug: When you are finished viewing TV for the night, unplug the set. You can also buy a line switch that can be spliced into your existing TV cord, but it must have a power rating equal to that of the set. The use of a regular extension cord containing a line switch or an inappropriately sized switch spliced into your TV cord is dangerous. If every American home had one color TV and each eliminated the instant-on feature, the annual savings would be hundreds of millions of dollars each year.

If you own a set with vacuum tubes, you should consider getting a solid-state model. It uses a lot less power. Similarly a black-and-white set costs only one third as much to operate as a color set. You may wish to consider one for your bedroom or for a child's room.

Dehumidifier

In regions of the United States where dampness is a problem (e.g., the Southeast and coastal areas), many families use a dehumidifier to dry out their basements. The typical owner of a dehumidifier uses it 1,465 hours a year. This involves about $35 worth of electrical power. During especially damp periods, a dehumidifier may be in use 8 to 12 hours per day, even though it might do a satisfactory job of removing moisture in just 1 or 2 hours per day. In fact, a savings of up to 75 percent on annual dehumidifying costs can often be achieved without changing your comfort level, simply by shutting off the dehumidifier sooner.

Humidifier

This appliance is normally used during the heating season to counteract dryness in the air due to space heating. A typical household that uses a humidifier spends about $15 a year to run it. The best way to reduce operating costs is to take advantage of alternative sources of moisture in the house. Venting an electric clothes dryer into the living space instead of to the outdoors can provide a great deal of moisture. Or use a clothes-drying rack instead of a power dryer. (Don't vent a gas clothes dryer indoors—it could cause indoor air pollution.)

Water-filled metal containers on radiators can also help. Steam cooking adds moisture and is energy efficient, especially if you use a pressure cooker. Plug up the tub when you take a shower and leave the water in for several hours. It provides both heat and moisture.

Coffee Maker

An automatic electric coffee-making appliance draws a large amount of power in order to brew your beverage quickly. If you turn it off as soon as your coffee is brewed, you'll save energy. Electric percolators use less power. A conventional coffee pot/percolator heated on a gas stove is the cheapest way to make coffee.

Toaster

A toaster draws a great deal of electricity, but the typical household only uses it 35 hours a year. Unless you make large quantities of toast, a four-slice model is wasteful. If you normally eat two slices, make both pieces of toast simultaneously rather than toasting single slices in sequence.

Electric Blanket

Heating your bed electrically is much less expensive than heating the entire bedroom. Turn down the room's thermostat and use a good-quality electric blanket to keep warm. Or try this time-tested approach: A hot water bottle under the covers 30 minutes before bedtime will make things very cozy. Add a few more blankets to help retain body heat. It's a natural way to stay warm. So throw on the blankets and cuddle up!

Deep Fryers, Electric Frypans, Roasters, Broilers, Hot Plates, Hot Dog Makers, Sandwich Grills, and Corn Poppers

Generally speaking, these essentially silly appliances are wasteful. They can save you money if you own an electric range: Using smaller appliances to cook some of the meals you normally prepare in your oven or broiler can be economical. But my suggestion is, if at all possible, convert to a natural gas range and use your stove-top burners. This will save the most money over the long run.

Electric Iron

Rarely does a family put in more than 2 hours a week ironing clothes. Synthetics have reduced the amount of ironing that people must do. So have T-shirts and jeans. Expect to spend $4 a year if you iron 1 hour a week, $8 if you iron for 2 hours. Turn the iron off 10 to 15 minutes before you finish and take advantage of reserve heat stored up in your iron. Items that normally require a low temperature are ideal for this coasting period. Folding clothes after drying reduces wrinkles. So does hanging them up. Remember also that steam works wonders on wrinkles. Hang a wrinkled item of clothing next to a steaming shower and notice the difference. You may find this a cheap and easy alternative to ironing.

Hair Dryer

This appliance uses very little electricity on a yearly basis. I personally believe natural air plus towel drying is healthier, but if you prefer to use a hair dryer, it will run you about $1 a year.

Stereo Player and Radio

Solid-state stereos and radios are not big energy users. Some savings are usually possible by merely remembering to turn them off when they're not in use. Often, a record album gets played, and it is hours before someone notices that the stereo is still on. Make a conscious effort to change this behavior—it could save you about 30 kwh a year. More if you have a teenager!

Sewing Machine, Clock, and Vacuum

These items do not consume very much power, especially because they are used on a limited basis. Using a carpet sweeper or sewing by hand are nonelectrical options, although in some situations they're not very practical. An electric clock costs about $1.50 a year to operate. Batteries may cost a bit less, but not necessarily. A windup pendulum clock costs nothing to operate, but its accuracy may not be great. Digital clocks are accurate and use little power.

Mixer and Blender

These are a good buy and are cheap and efficient to operate. Hand-mixing is good for your heart, although many people find it too laborious and time-consuming.

Waste Disposer and Trash Compactor

These devices typically use a combined total of about 80 kwh a year. The economic argument against them is not as strong as the ecological one. If all your organic scraps (e.g., coffee grounds, egg shells, orange peels, corn cobs, etc.) were used for garden compost, you would reduce the generation of household waste, ultimately save on garbage pickup costs, have less chemicals in your garden and water supply, and grow better vegetables and flowers.

Electric Water Pump

Most homeowners do not have wells, but many—especially in rural areas—do. A water pump will cost you $36 a year to operate if you average 300 gallons of water per day. Many families exceed this amount, especially if they use lawn and garden sprinklers. To save on water use (and hence pumping), follow the recommendations in chapter 2 on saving hot water and install flow restrictors in selected cold water taps. There are tank inserts for your toilet that reduce the amount of water per flush. I'd recommend instead that you merely place a few bricks in the tank and, if the flush is acceptable, try one or two more. Don't flush after every use. No, this won't make you feel like you're in the Third World or a New York subway men's room. You'll acclimatize very quickly.

Take a shower instead of a bath unless you normally take very long showers. Use short-cycle options on dishwashers and water recycling options on clothes washers, if you have these features. Mulch your garden—it holds water in the ground longer. Fix leaky faucets—they can use a lot of water. These measures could reduce overall water consumption and pumping costs by 30 percent.

Replacing Appliances

Most household appliances have a life expectancy of 10 to 15 years. You can therefore anticipate having to replace two-thirds of your major appliances over the next decade. When I recommend that you replace your existing electrical equipment with higher-efficiency models or with appliances that operate with an alternative fuel source such as natural gas, I assume that you will do this as your present aging stock is retired from use. If you now pay 10¢ to 15¢ per kwh for electricity and/or you are a heavy user of a particular appliance, you might wish to consider its early retirement. This is especially important when considering the potential savings of converting to efficient refrigerators, air conditioners, and clothes dryers. All three draw large quantities of electric power and thus savings can mount up quickly on high-efficiency units or with the switch to a cheaper fuel.

Table 8-1 gives the average life expectancy of household electric appliances. If you are considering replacement of equipment, use this information to roughly estimate how much longer you can expect each appliance to last.

Heat Exchanger

Throughout this book, I have advised switching to energy-efficient natural gas water heaters, furnaces, ranges, and dryers when you replace existing electrical equipment. In this way you will benefit from the double savings of energy-conserving technology and a fuel source that delivers three times as many Btu per dollar as electricity when burned in appliances that are 80 percent efficient. Despite its economic drawbacks, the one virtue that electricity does

Table 8-1
Average Life Expectancy of Electric Appliances

Appliance	Average Life (yr.)	Appliance	Average Life (yr.)
Air conditioner (central)	13	Hot plate	15
Air conditioner (room)	12	Iron	7
Blender	10	Microwave oven	15
Clothes dryer	12	Mixer	10
Clothes washer	12	Radio	12
Coffee maker	3	Range	14
Crockpot	10	Refrigerator	15
Dehumidifier	10	Sewing machine	24
Dishwasher	11	Television (black & white)	12
Fan	18	Television (color)	11
Freezer	17	Toaster	12
Frying pan	17	Vacuum cleaner	16
Hair dryer (blower)	2	Water heater	11
Heater (portable)	10		

exhibit is that it is a clean fuel. At its source of generation—the power plant—it is far from nonpolluting. But it *is* clean at your home. You don't burn electricity—no nitrous oxide or other residual by-products of combustion go up your chimney. Better yet, no combustion by-products pollute the air inside your home. You just flick a switch and receive clean electric power. Burning gas or any other fuel in the home, by comparison, does pose the potential risk of some indoor air pollution.

In a house that has not been caulked and weather-stripped, there is less of the health hazard associated with fuel combustion than is the case in well-insulated, "tight" homes. Tight houses are also prone to exhibit high levels of formaldehyde, radon gas, carbon monoxide, carbon dioxide, and smoke. A solution for this problem is the use of an air-to-air heat exchanger. It's sole purpose is getting stale inside air out and fresh outside air into your home, without major heat loss. The exchanger removes heat from indoor air that it vents to the outdoors, and it transfers this heat to fresh outdoor air that it pulls into the house. As much as 80 percent of the heating energy in the vented air is recovered. At $200 for an air-to-air heat exchanger, you can control the air quality in your home. The alternative—allowing your home to naturally exchange air by not sealing holes and seams—can result in huge heat leakages on windy days and inside air pollution on still days. An air-to-air heat exchanger will use $15 to $40 worth of electricity during the heating season.

Chapter 9
Conclusion

As we have seen, typical households should be able to reduce their electric bills by 50 percent easily. Families with electric space heating, water heating, and ranges ought to be able to save much more—up to 80 percent. The "all-electric" household thus might reduce its total energy bill to only a fifth of what it is now. The first 50 percent cut should be both simple and cheap to carry out through the adoption of conservation measures. Greater savings come by shifting to alternative fuels and new high-efficiency appliances. These appliances involve a higher initial investment than conservation, so you may have to borrow the money, but with electricity costs climbing, you'll begin to save significant amounts immediately—even while you're repaying the loan.

If you proceed with conservation retrofits and the purchase of new equipment, make sure you are an educated consumer. Look for high-quality, energy-efficient products. *Consumer Reports* is one of the best places to go for quick, reliable information. It will tell you which refrigerators, dishwashers, heat pumps, solar hot water systems, air conditioners, thermostat control devices, furnaces, and so on, will truly give you the most output per dollar spent in operating cost. *New Shelter* also publishes product evaluations that are based on objective testing; these can be of great help when you are selecting items to purchase. For additional information on many products, see part 2 of this book.

Table 9-1 shows you the effect that following my suggestions will have on your overall electric bill. Generally speaking, achieving a 25 percent drop in consumption should cost you around $100, a 50 percent reduction will run around $500.

The conservation measures cited in table 9-1 can bring you a 43 percent reduction in power use. Over time, conversion to high-efficiency appliances and shifting to other fuels can just about double this figure, bringing you an ultimate reduction of 70 to 80 percent. Beware, however, that the shift to some energy-efficient appliances can potentially hook you into more electrical consumption. For example, most electric companies will do an energy audit of your home for free. One of their representatives will analyze your current energy situation, but he or she may not be completely objective in his or her recommendations. If you now own inefficient oil or gas equipment, do not allow yourself to be persuaded that a new electric heat pump or some other electrical appliance is necessarily the right choice for you. You may save a bit of money on the cost

Table 9-1

Financial Impact of Conservation Efforts*

Task	Before kwhs per Month	Before Cost per Month ($)	Reduction (%)	After Cost per Month ($)	After Reduction Method and Cost†
Water heating	530	47.70	50	23.85	Insulation, timer, flow restrictors: $70
Lighting	100	9.00	38	5.58	Conversion to more efficient bulbs, use of a dimmer: $15
Refrigeration	153	13.77	25	10.33	Replace gasket: $5
Cooking	96	8.64	50	4.32	Purchase supplemental appliances: $100
Room air conditioning‡	117	10.53	40	6.35	Attic insulation, timer, use of high vent, shading: $200
Clothes drying	91	8.19	50	4.10	Use clothesline: $2
Dishwashing	30	2.70	50	1.35	Skip automatic drying cycle, wash dishes by hand sometimes: $0
Running color T.V.	36	3.24	40	1.94	Pull plug: $0
Running miscellaneous appliances	95	8.55	30	5.99	Timer, odds and ends: $30
Total	1,248	112.32	43	63.81	

Cost of Reductions: $422
Annual Savings: $582

*Electric rate: 9¢ per kwh.
†Some additional no-cost conservation measures carried out but not listed.
‡10,000 Btu, EER 7, operated for 1,000 hours a year.

of operating your present appliances, but not as much as you could save with efficient nonelectrical alternatives.

Be equally skeptical of advice to purchase an active-solar hot water system. Your energy auditor may recommend that you convert from an old gas-fired hot water system to a solar water-heating system. This could be a mistake in some circumstances. The sun does not shine all the time, and when it is not shining, most active-solar systems use electricity to heat the water. Moreover, unless a solar water-heating system is properly designed, it will operate inefficiently. Perhaps most important, you must be sure that the solar collectors are properly sized for the size of the system's hot water tank. A large 80- to 120-gallon water tank connected to undersized solar collectors will not be able to heat your water to an adequate temperature most of the time—even in summer. This means your solar water system will become a permanent source of demand for utility power, and it will use a great deal of it at an ever-increasing cost. If you do opt for a solar water-heating system, be sure you have sufficient collector area and look for a system with nonelectric backup. My general advice is to consider electrical appliances only as a last resort. Go to any other fuel source before you go to electricity.

Alternatives to Your Electric Company

There are utility companies around the country that have turned away from the notion of building new power plants to meet the demand for electricity in their regions. Instead, they are encouraging their customers to reduce demand through conservation and are in fact providing low-interest loans to help them make these changes. This means that instead of generating new kilowatts, these utilities conserve power at a cost as low as one-twentieth the amount needed to construct new power plants. These utilities are also purchasing power generated by solar, wind, and hydroelectric power equipment—a far wiser bet than building new nuclear or fossil-fuel plants.

Soon homeowners will also be able to take advantage of renewable energy sources and will generate their own electric power using the sun, the wind, and water. Photovoltaic solar panels, for example, generate electricity through the use of solar cells. They are reliable and environmentally safe. They are presently expensive—50¢ to $1 per kwh over their average life of 20 years. They are dropping in cost, however, even faster than electric rates are climbing. In fact, it is now expected that in the near future solar panels will produce power at 10¢ to 15¢ per kwh. In some areas of the United States, utility-generated electricity already costs this much. Thus, photovoltaic systems will soon be quite cost-competitive as a source of electricity.

Some people will generate their own electrical power and store enough for home use in batteries to cover sunless periods. These people will cut their dependency on the utilities forever. Others will generate a surplus and sell it back to their local utility. A federal law requires that utility companies buy surplus power from individuals at a fair price. Many people will trade their surplus power during sunny periods for supplemental power during periods when they are not generating enough electricity to meet their needs.

Conservation-conscious households can get along well on 300 kwh per month or less, as long as they do not use electric space heaters, water heaters, or ranges. In the early 1990s, by spending approximately $7,000 for photovoltaic panels and auxiliary equipment, you should be able to assure yourself of at least 20 years' worth of electricity that will be cheaper than the power sold by utility companies.

If you live in an area with a moderate or better-than-average wind speed (i.e., 12 mph and above), you may wish to look into home wind electrical generation systems, which are expected to cost 8¢ to 12¢ per kwh by the early 1990s. Small-scale hydroelectric systems that produce electricity by harnessing streams are also a possibility for people who have access to a small creek or stream on their land. Prices are currently 5¢ to 18¢ per kwh.

The path of liberation is clear. Implement as much conservation as you possibly can as quickly as you can, even if you must borrow the money to do it. Convert to energy-efficient nonelectric appliances whenever possible and to energy-efficient electric ones only when alternatives are not available or practical.

For more information about generating your own electricity, see the section of part 2 titled "Alternative Generation of Electricity."

Special Cases

Let's close by considering the steps you can take if you find yourself in one of the following special situations:

If You Rent

Many people choose to rent, and millions of others are forced to rent by high interest rates. If you are a tenant, you face a special dilemma in that energy-saving modifications to your dwelling are frequently not economically justifiable. At the same time, landlords may not care if tenants have to pay high electric bills.

Generally speaking, two types of electricity-saving methods are available to renters. Practicing good conservation habits is the first. This approach involves behavior modification. It is the cheapest form of cost cutting—usually free—but it is also the hardest to act on because you must consciously and actively change your habits of consumption. My tips on hot water use, cooking, refrigerator use, lighting, and so on, apply here. These conservation tips will save you money, but you must actively participate in the daily process of using electric power conscientiously.

The second approach available to renters is that of carrying out no-cost or

low-cost modifications to household appliances. These probably have the highest potential for saving money because they involve changes that can be carried out and then forgotten. Here we can include turning down the temperature on hot water and heating equipment (no cost); turning up the temperature on refrigerators and air conditioners (no cost); installing water tank insulation and flow restrictors (low cost); replacing seals on refrigerators, freezers, and oven doors (low cost); switching to more efficient light bulbs and dimmers (low cost); unplugging TVs or using TV on/off power switches (low cost); caulking and weather-stripping around windows and doors (low cost); using insulating shutters or shades over windows in winter (low to moderate cost); and using shades or blinds over windows in summer (low to moderate cost).

Renters should also consider purchasing some of their own household appliances, especially if electric rates in their area are over 9¢ per kwh. Refrigerators, room air conditioners, freezers, fans, dishwashers, and clothes washers can be big energy users, and some newer models operate at half the cost of older ones. These appliances belong to you, so you can take them with you when you move. The same applies to any air conditioner precooling misters and timing devices for water heaters, space heaters, and air conditioners that you buy.

The area of savings with the lowest potential for renters is making modifications to the dwelling or conversion to new fixed equipment (i.e., equipment such as water heaters that stay with the apartment when you move). Tenants could not be expected to invest in these changes and landlords probably will perceive no direct economic advantage in making them. Some negotiation may be possible, however. Insulation, caulk, and weather stripping have become a human necessity in this age of high fuel costs, so individuals or groups of tenants should pressure landlords to invest in such measures. Wasting energy has become un-American. Perhaps if all other pleas fail, you could offer to install the insulation, etc., if your landlord buys the materials.

When appliances belonging to the landlord need replacement, you might approach him or her about purchasing more energy-efficient models. Usually there is a relationship between the high-efficiency level of an appliance and its overall quality. Better equipment will cost your landlord less in maintenance over the long run.

Generally speaking, the major opportunities for renters to reduce electricity use cost little or nothing, but can still put a big dent in their electric bills. Up to a 40 percent savings should be achievable.

If Your Bill Seems Too High

Check your bill. The majority of residential customers come under a general rate. Electric companies sometimes offer reduced special rates to customers with electric hot water systems or electric baseboard resistance heating systems.

Your rate code should be shown on your bill. Check your rate code and make sure you are being charged the proper rate.

Rates may fluctuate seasonally or even daily. Many utilities raise rates during the summer when air conditioners are used. Companies may also charge a lower nighttime rate just like the phone company does. Higher summer and day rates discourage unnecessary consumption during these "peak demand" times. This is actually a sensible policy, because it motivates people to redistribute their use of electric power. If people were not given incentives to curtail summer and daytime use, utilities would have to add new power plants to meet increased demand for electricity, despite the fact that these plants would not be needed during the winter and at night.

Thus, when checking your bill, be aware that rates vary according to both the type of equipment you have and the times of the day or year you are drawing power. Make sure that your bill accurately reflects the rate you should be paying. Look carefully at the possibilities for an alternative discount rate if in fact you can qualify.

The "service period" is the time from your previous meter reading to the most recent one recorded on your bill. This time period should be normally either 30 or 60 days. The number of kwh you used during the service period multiplied by your rate per kwh will give you the amount you are being charged for electricity over the service period. Thus, if you are charged at a rate of 9¢ per kwh and you used 500 kwh during a 30-day period, your bill should be $.09 × 500 or $45. Add any special charges listed on your bill to determine the final figure. Special charges may include a flat service charge of several dollars and any sales taxes that are applied. If your final figure does not equal the amount of your electric bill, contact your utility about the discrepancy.

If you have some doubt about the meter reading itself, contact your electric company to determine whether your last reading was an actual reading or an estimate. If it was an estimate, it may not reflect your actual consumption over the service period. If it was an actual reading and you have doubts about its accuracy, check the meter yourself. A quick, although somewhat unreliable, method is to do a 10-day reading yourself. For example, if at the start of the 10-day period, your meter reads 46,600, and at the end of the period it reads 46,800, you have used 200 kwh in 10 days or 20 kwh daily. If your previous bill indicates that you used between 500 and 700 kwh over a 30-day period, then the bill also reflects a use of approximately 20 kwh daily, so the company's reading is probably accurate. But if the company's figure is 800 or 900 kwh for 30 days—about 26 to 30 kwh daily—you may want to do some further checking. Before you do, however, think carefully about any extra use that may have occurred in the previous service period. Did you have house guests, fill a swimming pool, or use any heavy-draw equipment when you had some work done on your home? Did you use your electric oven more than usual? If you can not come up with a logical explanation, you might consider getting an electrician to

check your meter. Experience has shown, however, that meters are rarely at fault.

Your Bill Is High while You Are Away

Several years ago, a family complained to me that they had been away for a month yet their electric bill for that time was almost as high as when they occupied the house. Thus, they argued, the electric company or the meter must be wrong. I did some simple checking around, however, and was able to explain why their bill remained high even in their absence.

They lived in New York State and had gone away in January—the coldest month. They had an oil-fired forced-air heating system and left the thermostat on 55°F to avoid problems with frozen pipes and also because there were plants in the house. Heating one's home to 55°F while you are away is actually a common practice. They expected to have a substantial heating bill to pay for the oil, but why did they have such a high electric bill?

To keep the house at 55°F meant not only use of the oil burner but also of its electric blower, which distributes heat through ducts. They also had an electric hot water system. It was set at 160°F, which is abnormally high to begin with. Because the house was kept at 55°F instead of the normal 68°F, heat losses from the water tank to the surrounding environment were huge. The water heater was located in the basement. Since the upstairs interior of the house was only 55°F, the basement temperature was probably between 45°F and 50°F. The temperature difference between the water and the basement air was therefore 110°F. The amount of electricity required to boost water temperature this much is phenomenal. Since they had not insulated the outside of their tank, the heat loss occurred very quickly and electricity was used by the water heater almost constantly.

They had enough forethought to disconnect their refrigerator, but it never occurred to them that their instant-on TVs were constantly using electricity. They had one color and one black-and-white TV. Both were left plugged in. Add to this several electric clocks that were left running, two 60-watt nighttime security lights, and a flat service charge of $3.60 per month in their area, and it was easy to see why they had such a high bill.

If you will be leaving your home for any period longer than a few days, I would recommend completely shutting down all systems. Almost every hot water and cold water system can be easily drained via a drain valve, usually located in the basement. Your neighbor or plumber should be able to tell you where it is if you can't locate it. If you open it and let it drain completely, all your pipes and your water tank will be emptied of water that could freeze. Be sure to shut off power to the water heater before draining it. Make sure you also shut off the valve that brings water into your house from your city water supply. If you have a well, pull the fuse or turn off the circuit breaker that operates your water pump. A tablespoon of alcohol in every drain should prevent any freezing that could occur in drain traps. Toilet tanks and bowls must also be emptied of water and you should stuff a rag in the toilet bowl to prevent combustible gases from entering the house while you are away.

If you wish to have a timed security light on, you should not cut power off completely. But unplug TVs, clocks, and all other devices that would use electricity in your absence. If you're willing to do without security lighting, just pull the main power switch. You should get the advice of a plumber or electrician or an expert neighbor on these procedures the first time around, just to make sure. After that, you will know how to do them.

So, we have now covered all the topics related to slashing your use of electricity in the house. By following the advice given earlier in this book, you can minimize electrical consumption when you are at home. And by following the recommendations in this final section, you can slash—or even eliminate altogether—electrical use while you are away. This energy conservation will prove its worth year-round. When you leave on a vacation, for example, you might take a portion of the money you've saved, and use it to enhance your holiday. Or you can put all of your savings in the bank, where it will accrue interest for you. Either approach sure beats handing your money over to the utility company.

PART 2

In this half of the book, I will present more detailed information about various types of energy-saving systems you may want to consider. Each section will conclude with a list of some manufacturers and the equipment they produce. Additional systems may be available to you locally—check your phone directory.

Most of the systems I will list have been recommended—or at least favorably discussed—in consumer-oriented magazines that focus on energy-efficient products. By contacting the manufacturer, you can receive the address of the nearest retailer.

Section 1

Energy-Efficient Water Heaters

Although there are new conventional electric water heaters that can give you more hot water per kilowatt-hour than the older models could, the savings are only marginal and I would not advise converting to such heaters. Electrically operated heat pump water heaters are a better bet, and new high-efficiency natural gas systems are better still: They can radically reduce your water-heating costs.

[1] Electrically powered air-to-water heat pumps are extremely efficient because they extract heat from the air rather than generating their own heat directly from electricity through a high-resistance coil. As is true of all heat pump technology, the ratio of heat energy captured to the electrical energy consumed drops as air temperatures go down. In the colder northern climates where heat pumps must be placed indoors to assure adequate temperatures for equipment operation (optimal range is 45°F to 95°F), winter use can steal some of your house's heat. If you use an expensive space-heating fuel like electricity, the heat pulled from indoor air by an indoor heat pump can significantly offset any savings you achieve on the cost of water heating.

There are two types of heat pump systems available: add-on units that connect to your existing hot water system, and all-in-one units that replace your existing system. Add-on units allow you to extract heat from the air during the warmer times of the year when heat pump efficiency is greatest, then they let you use your existing system for most of your hot water during colder periods. The all-in-one units contain their own resistance coils to heat water in a conventional manner whenever temperatures prevent the heat pump from extracting enough heat efficiently from the air.

Generally speaking, heat pumps have a slower recovery time than other systems (i.e., it takes longer for them to heat a given quantity of cold water). Some systems compensate for this sluggishness by using resistance coils. This approach can cut into your savings. Despite these disadvantages, though, a savings of 50 to 60 percent compared to conventional electric water-heating costs is typical. The higher the efficiency of the pump—as indicated by its coefficient of performance (COP) rating—the greater your savings are likely to be.

2 New gas-fueled hot water systems are an especially desirable alternative to electric units. Older and less efficient gas models with the burner beneath the storage tank lost a great deal of heat to the area below and around the flame. Efficient new models have a submerged heating chamber within the tank, so heat losses are much lower. Heavy built-in tank insulation has further reduced heat losses. Gas water heaters have a faster recovery time than either conventional electric or heat pump units.

When selecting a gas-fueled water heater, look for "regulated power combustion," which gives a better fuel/air mixture and a more efficient burn. With an electronic ignition, you will not need a constantly burning pilot light and this will also mean fuel savings. Most models require a flue extending through the roof to vent combusted gases. New direct-vent models can eliminate the need for flue pipes with the use of a through-the-wall vent.

Note that water-heating systems can be made more cost-effective by reducing water temperature settings at the tank and using timers to control on/off cycles. Low-flow shower heads and faucet flow restrictors can also reduce your hot water load. There are two general types of low-flow heads: fine-spray models and aerating heads. The former use smaller holes or pores to restrict water passage, while aerators mix air and water to produce a fine cone-shaped spray. Look for shower heads (of either type) with flow rates of 2 gallons per minute or less. Look for faucet aerators with flow ratings of less than 3 gallons per minute. These can save you up to 60 percent at the tap.

Keep in mind that the conversion to a gas point-of-demand water heater eliminates "standby" heat losses completely (these are losses that occur when a tank full of hot water cools off; a point-of-demand heater has no tank, hence no standby losses). Point-of-demand heaters are often the best option. See the following section, "Tankless Gas Water Heaters."

Gas Water Heaters

Hot Water Maker HP-40A
Amtrol, Inc.
1400 Division Rd.
West Warwick, RI 02893
(41-gal. storage; 77 gal. per hr.;
64 percent efficiency.)

Rheem Glas Fury Energy Miser Series
Rheem Manufacturing Co.
5780 Peachtree, Suite 400
Dunwoody Rd.
Atlanta, GA 30342
(30-, 40-, and 50-gal. storage; 50 gal.
per hr. on 30-gal. tank; 61 percent
efficiency.)

Heat Pump Water Heaters

E-Tech Efficiency II, 100 Series
E-Tech Corp.
2115 American Industry Way
Atlanta, GA 30341
(COP 2.2 to 3.0; 17 gal. per hr. recovery
to 135°F; add-on unit.)

Therma-Stor HP-52, HP-80
Therma-Stor Products Group
P.O. Box 8050
Madison, WI 53708
(COP 3 to 3.4; 52- and 80-gal. storage;
stand-alone units.)

Flow Restrictors

American Standard Astro Jet Shower
Head 1413111
American Standard Company
P.O. Box 2003
New Brunswick, NJ 08903
(1.3 gal. per min.; 80 percent reduction.)

The Incredible Head Elite ES271 Shower
Head
The Incredible Tapsaver
Resources Conservation, Inc.
P.O. Box 71
Greenwich, CT 06836
(Shower head flow: 1.9 gal. per min.;
70 percent reduction. Tapsaver flow:
2.75 gal. per min.; 60 percent reduction.)

Moen Easy Clean 3900A Shower Head
Stanadyne Corp., Moen Div.
P.O. Box 4007
377 Woodland Ave.
Elyria, OH 44036
(2.4 gal. per min.; 60 percent reduction.)

Power Flow, PF 20 Shower Head
Viatek Industries, Inc.
1730 E. Prospect St.
Fort Collins, CO 80525
(2 gal. per min.; 70 percent reduction.)

Section 2
Tankless Gas
Water Heaters

The major rationale for converting to a point-of-demand tankless water heater is that such a heater eliminates "standby losses" (i.e., conductive heat losses from the hot water in the tank to the surrounding environment). This is an important objective. To fully understand your options, however, you need to know a little background information.

Standby losses rarely constitute more than 20 to 25 percent of your total hot water costs. If your present water heater is inside your heated living space and/or the tank is moderately to well insulated, these losses will be much less. You can insulate an existing tank to bring standby losses to a minimum.

If you currently have an electric water heater and have carried out basic conservation procedures, do not expect to save much more than 20 to 25 percent on your monthly hot water bill by installing an electric tankless system. On the other hand, the installation of a *gas* tankless system as an alternative to your existing conventional electric water heater is a totally different story.

Gas tankless systems use the same kind of heating elements or burners as conventional gas water heaters. The difference is that conventional systems use energy whenever tank water temperatures drop below a preset level. This usually happens when you draw hot water out of the tank and cold water takes its place, or over time when a tank of heated water gradually loses energy through tank walls to the surrounding air (the dreaded standby losses).

Tankless units, by comparison, do not heat any water until you need it. They respond to water flow. When a faucet or shower is turned on, sensors activate a tankless unit, which heats up cold water on the spot. To heat 50°F incoming well or city water quickly, as these units do, requires a great deal of energy at one time—up to 125,000 Btu per hour versus 15,000 Btu per hour for conventional systems. But since this energy is only drawn when water is being used, and not to maintain a 60-gallon tank at 120°F to 160°F for hours on end, the overall effect is lower total energy consumption.

By switching to a tankless gas-heated system, you will save due to the elimination of standby losses—and you will save even more for the simple reason that you have changed from electricity to gas. The typical American family pays around $48 a month for electric hot water; they would experience a drop to $16.50 a month using a 75 percent efficient gas tankless system.

The distinct disadvantage of tankless systems is that they can't produce as much hot water as fast as conventional systems can. The latter can provide

many gallons of hot water a minute for a limited period of time whereas tankless systems rarely can give you more than 2½ gallons of 120°F to 130°F water per minute. A tankless system can, however, give you this amount of hot water indefinitely. Moreover, by spreading out your hot water needs over different times of the day, the relative slowness of tankless heaters can be overcome.

Older tankless models gave a fixed heat output and thus the temperature of the water varied according to how fast it flowed (i.e., you could get less water at a higher temperature or more water at a lower temperature). Newer "modulating" systems allow integration of temperature and flow settings so that you may choose a desired temperature within limits. Most models also contain an automatic cut-off switch so that slow-flowing water does not overheat.

Gas units have a piezoelectric or spark pilot-light ignition. Although it is common to leave the pilot light burning all the time, you may decide to turn off the pilot for some or most of the time and relight it whenever you wish to use it. This is not as inconvenient as it sounds since tankless water heaters are usually installed in closets, bathrooms, or kitchens where they are accessible. Turning off the pilot will, of course, save additional energy.

Gas tankless heaters are easy to install, but for safety's sake the work should be done by your gas company or a plumber. Although most gas units require a flue, direct-vent models are available. These allow you to use outside air for combustion and they exhaust their waste gases directly through an exterior wall. Natural-gas systems can use liquid propane gas (LP), which is more expensive than natural gas but less expensive than electricity in most areas. The largest tankless units can meet the overall hot water demands of a family, whereas smaller units are ideal for particular tasks such as boosting water temperatures for a dishwasher.

Large Tankless Water Heaters

Aqua Star 125 VP
Controlled Energy Corp.
Box 19, Fiddler's Green
Waitsfield, VT 05673
(2.2-gal. flow at 130°F; 125,000-Btu-per-hr. capacity.)

Paloma Constant-Flo PH12M-DN
Paloma Industries, Inc.
241 James St.
Bensenville, IL 60106
(2.2-gal. flow at 130°F; 89,300-Btu-per-hr. capacity.)

Thermar TL 100, Homemaster DV100
Thermar Tankless Heater Co.
41 Monroe Turnpike
Trumbull, CT 06611-0398
(2.3-gal. flow at 130°F; 100,000-Btu-per-hr. capacity; direct-venting unit requires no chimney.)

Small Tankless Water Heaters

Instant-Flow S-60C/22OU
Chronomite Labs
21011 S. Figueroa St.
Carson City, CA 90745
(.5-gal. flow at 140°F; 20,300-Btu-per-hr. capacity; electric unit for boosting dishwasher temperatures.)

Stanton HW 300
Thermar Tankless Heater Co.
Therma Center
Trumbull, CT 06611-0398
(1- to 1.5-gal. flow at 105°F; wall-mounted TAP-type electric booster unit.)

Section 3
Residential Solar Water Heaters

Solar hot water systems almost never meet the full water-heating needs of a household, but they typically do provide 30 to 80 percent of those needs, with the remainder provided by a conventional water heater, usually one that uses electricity. The amount of hot water you get from a solar water-heating system depends on a number of factors, including the amount of sunshine in your area; the overall efficiency of the system you select; the size, tilt angle, and orientation of the collectors; and the daily and seasonal average temperatures in your region. A very generalized rule of thumb is that on a sunny day, each square foot of collector area will give you around a gallon of hot water in spring and fall, more in summer, less in winter. A family of four would need 40 to 80 gallons a day depending on their habits and adherence to water-conservation techniques.

Residential solar hot water systems can be either active or passive. The former use pumps and other moving parts to force-circulate water or an antifreeze solution between the collectors and the storage tank. The advantages of active systems include their capability to operate in cold weather and to meet a higher proportion of a household's hot water needs. They are, however, more complex and more expensive than passive systems. Passive systems contain no moving parts and operate via natural energy flows.

Many solar water-heating systems use "flat-plate" collectors, which are thin insulated boxes that are usually mounted on the roof of a house. Each flat-plate collector contains a metal absorber plate painted black, since dark surfaces absorb and hence capture a high proportion of the solar energy that strikes them. The collector box is insulated at the back and sides, and is double glazed with a transparent sheet of rigid plastic or glass on the front. The light that passes through the glazing is converted to heat when it strikes the absorber plate. This heat is in turn conducted to water (or an antifreeze solution) that travels through pipes or tubes (called risers) that are bonded to the metal plate.

The bonding between the pipes and the plate is the primary path of heat transfer from the plate to the fluid. The fluid then travels to a water-storage tank, which is usually located inside the house.

In cold climates where freeze protection is necessary, antifreeze can be pumped in a closed loop between the collectors and the storage tank. The antifreeze solution doesn't come directly into contact with the water that will come out of your water taps. Instead, a heat-exchange device—usually a coil with substantial surface area—is placed inside the water-storage tank. The antifreeze solution circulates between the collectors and the heat exchanger. Thanks to the heat exchanger, the antifreeze fluid gives up much of its heat to the water that fills the storage tank.

Another configuration uses regular household water in the tank-collector loop instead of antifreeze. In such a "drainback" system, water automatically drains out of the collectors and back into the storage tank whenever the pumps are not running. A thermostat controls the pump and cuts off its operation whenever there is danger of water freezing in the collectors.

A kindred system—the open-loop system—prevents freezing at the collectors by recirculating warm water in the storage tank through the collectors when outside temperatures approach freezing. This approach is ideal for areas of the country with moderate climates where temperatures drop to 32°F only a few times a year.

The oldest and most dependable solar water-heating system is the passive thermosiphoning or free-flow model. It uses no moving parts and does not require electrical power to operate. It consists of several collectors mounted on a house or greenhouse roof and a hot water tank that must be positioned 18 to 24 inches above the collectors. Natural convection—the tendency of warm water to rise—occurs when water in the collector is warmed by the sun. The water rises to the storage tank and mixes with the water already inside the tank. Over time, the overall temperature of the water inside the tank increases, with the hottest water rising to the top of the tank where it can be drawn off for use. Cooler water at the bottom of the tank sinks into the collectors for heating. This thermosiphon system is ideal for warm climates and for summer use. Since it makes no provision for freeze protection, the system is normally drained in colder regions during the winter months.

The "orientation" of a collector is critical. Solar collectors should face south (i.e., be oriented toward the south) because, in the northern hemisphere, the sun is always in the southern sky: Its path is from the southeast to the southwest. The "tilt angle" of the collector is also important—it is the angle between the collector and the horizon (i.e., a collector that is positioned vertically has a tilt angle of 90 degrees; one that lies flat has a tilt angle of 0 degrees). The tilt angle affects a collector's efficiency because the more directly the sun's rays are intercepted, the higher the energy gain per square foot of collector surface. Generally, a tilt angle equal to your latitude is considered best. Thus, if you live at 40 degrees north latitude, the desired tilt angle for your collectors is 40 degrees.

Some solar water systems do not use flat-plate collectors. The batch water heater (sometimes called a breadbox heater) is an example. The storage tank itself is painted black and mounted in an insulated box with south-side double glazing. A typical configuration consists of one or two such units using 30- to 50-gallon tanks. Batch heaters can be mounted on the roof or on the ground. They are spliced into the cold water pipes between the incoming water supply and the existing water heater. They thus become a solar preheating system rather than a main source of hot water. Batch heaters must be drained during periods of the year when freezing temperatures occur.

A batch system with two 40-gallon tanks, double glazing, and insulating shutters used over the glazing from 5:00 P.M. to 8:00 A.M. can meet around 50 percent of the annual hot water needs of a typical family in the warmer regions of the nation. In regions that are cold but sunny, the same batch system should provide one-third of the annual water-heating load. In other regions, it should meet at least 25 percent of the load.

Passive systems tend to be more efficient than active ones, but they are generally smaller, so their contribution to the hot water load is generally lower. The area of collector surface tends to be about twice as large for active systems as for passive (e.g., 60 square feet versus 30 square feet). An active system may provide 50 to 80 percent of a family's hot water.

A batch system can be installed by a do-it-yourselfer for $300 to $1,500, and by a contractor for $1,500 to $2,500. Thermosiphoning models cost $1,000 to $1,500 if owner installed, or $2,500 to $3,000 otherwise. Active systems range in cost from around $1,500 up to $6,000.

In order to meet 75 percent of the annual hot water requirements of a family of four with a flat-plate system in the northern half of the United States, you will need anywhere from 60 to 90 square feet of collector area, whereas southern households will need between 30 to 60 square feet. Equipment to be used in the warmest regions will require only minimal freeze protection. These factors mean that residential solar water-heating systems can be especially valuable in the South, especially where electric rates are high. In the North, the cost of additional collector area and freeze protection make domestic solar hot water less of a bargain. Where electric rates are high, however—say 10¢ per kwh—60 to 70 square feet of roof collectors can save a typical family about $475 a year and give a payback period of seven or eight years.

Active-Solar Water-Heating Systems

Eagle Sun Systems
U.S. Solar Corp.
P.O. Drawer K
Hampton, FL 32044
(Open-loop, closed-loop, drainback, and thermosiphon systems; 66- and 120-gal. storage; electric backup.)

HWSGF 120/80 DD
Mar Flo Industries, Inc.
18450 S. Milles Rd.
Cleveland, OH 44128
(Draindown system; 120- or 80-gal. storage.)

The Hydrocratic
American Sunsystems, Inc.
829 First Ave.
P.O. Box 227
West Haven, CT 06516
(Draindown system; integrates with your existing tank.)

Solarcraft
State Industries, Inc.
By-Pass Rd.
Ashland City, TN 37015
(Drainback system; 82- or 120-gal. storage; electric booster.)

Passive-Solar Water-Heating Systems

Cornell 480
Cornell Energy, Inc.
32 N. Stone, Suite 901
Tucson, AZ 85701
(Batch system; 42-gal. storage.)

Solon 240, 120, 170
Amcor Group
7946 Alabama St.
Canoga Park, CA 91304
(Thermosiphon systems; 64-, 32-, and 45-gal. storage.)

SS 1,000
Sun Systems, Inc.
3831 E. Broadway
Phoenix, AZ 85040
(Batch system; 34-gal. storage.)

Section 4
Windows, Skylights, and Related Devices

Windows and skylights allow natural daylight to enter our homes. Unfortunately, they can also inflate our electric bills by letting heat escape during the winter and by letting hot sunlight in during the summer. A single-pane window will lose ten to fifteen times as much heat energy on a cold night as an equivalent area of minimally insulated wall. In summer, when outside temperatures are 75°F to 80°F, a home's interior can reach over 90°F due to solar heat entering through windows.

For years, heat losses were combatted by the installation of exterior storm windows. This was only minimally effective. In the era of cheap energy—up through the mid-1970s—exterior storm windows were adequate for most householders. But increases in the price of electricity have brought a new concern with windows, and in the last decade new materials and technologies have emerged.

Multiple-pane windows have become a popular alternative to single-pane windows equipped with exterior storm windows. Single-pane windows have an R-value of .9. Double-pane windows have two layers of glass and a dead-air space of from ⅜ to 1 inch between them. Such windows have an R-value of approximately 1.7. In the North, some people have gone further, turning to windows with three layers of glass, yielding an R-value of about 2.8.

Each layer of glazing that is added to a window brings a further increase in R-value. There is a problem, however. Besides cutting heat losses, additional layers of glass cut the amount of warming sunlight that can pass through a window. Single-, double-, and triple-glazed windows respectively allow 85, 72, and 61 percent of the available sunlight into the home. Clearly, if you want to benefit from solar warmth, there's a limit to how many layers of glazing you will want.

A partial solution is to install windows having low-iron glass. This glass offers the same R-values as regular glass, but it reduces incoming sunlight by only 6 to 8 percent for each glazing layer compared to 15 percent per pane for regular glass.

Another approach is to use windows that have layers of high-transparency plastic films between the panes of glass. These windows offer lower cost and lighter weight due to the use of their films, and they allow as much light through as low-iron glass. Thus, a "quadpane" window (one having four layers of glazing) would have an R-value of 3.8 and still allow 63 percent of available solar energy through. According to one manufacturer, a quadpane window that faces south would provide the equivalent of 47 kwh per square foot during a Minnesota winter. An east- or west-facing window would capture 21 kwh.

A less expensive option is the use of interior storm windows that you install yourself. These are made of clear plastic films or rigid acrylic. If they fit tightly, they can boost window R-values and seal against cold-air infiltration. Interior storms are cheaper than exterior storms—some sell for as little as 12¢ per square foot.

Movable insulation is another popular and effective means of fighting window heat loss. Normally used with double-glazed windows or single-glazed windows having storms, movable insulation (i.e., insulating interior shutters, quilts, drapes, shades, and curtains) allows sunlight in during the day and then radically curtails nighttime heat losses. Although some types of movable insulation are motorized, most require human energy to open and close them at the proper times. Prices range from $4 to $30 or more per square foot, so the payback period at the top end of this range can be quite long. On the other hand, do-it-yourself kits and the cheaper systems priced at $4 to $8 per square foot offer faster returns on your investment.

Interior insulating shutters normally contain a core of foil-covered rigid insulation encased in a lightweight wood exterior. R-values can be 6 or higher. Nonrigid interior insulating shades, curtains, blankets, and quilts generally provide a dead-air space between themselves and the glazing, and they use some combination of fiberfill and reflective foil in a fabric covering. Most are manually operated and form a tight seal around windows using tracks, channels, snaps, magnets, or Velcro.

Another option is to install coated plastic window films on your existing windows. The films are coated with dyes or metals and operate selectively on the passage of solar radiation, interior radiant heat, and visible light. Some are primarily designed to reduce heat loss by reflecting radiant interior heat back into the house, while others—ideal for hot climates—block unwanted solar heat by reflecting sunlight away from the window. In both cases, up to 80 percent reflection is possible compared to 15 percent for conventional glass. Note that coatings designed to reduce window heat losses will also reduce incoming solar heat to some degree.

Films are reasonably priced at 50¢ to $3 per square foot. Films that emphasize interior heat reflection are known as low-emittance or low-e films; look for one with a high R-value. If you purchase a sun-control film, look for a low shading coefficient rating—the closer to zero, the less solar heat will penetrate. Since window films also reduce incoming light, there is a trade-off: a high shading coefficient can mean you will be looking out through darker glass. In fact, a third purpose for using films is to reduce glare.

A recommended approach in hot climates is the use of traditional low-tech devices. These include sun screens that reduce incoming light by up to 80 percent while still permitting air flow; roll-up canvas shade cloths made of light cotton that allow diffuse light into your room while blocking 60 percent of incoming solar radiation; and the old standbys: roll-up shades and venetian blinds. The latter are now made with high-reflection surfaces and can be especially effective.

Skylights and roof windows (skylights that rotate 180 degrees for easy cleaning) present unique energy-related virtues and liabilities. On the positive side, a skylight can provide substantially more light to a room than an equivalent regular window, thus reducing the need for electric lighting. South-facing units can provide solar heat, although roof slants of 50 degrees or less mean you'll get more solar heat in fall and spring than in winter, when the sun is lower in the sky. A third energy advantage, perhaps the most important, is that openable models provide a high vent that is ideal for exhausting hot interior air in summer.

On the negative side, skylights can give you excessive solar gain in summer when you least want it. In winter, you have the worst sort of heat-losing surface— a window—in the greatest heat loss location in your house—the roof. To deal with these problems, you should use some of the strategies discussed above. Sun-control films, sun shades, and blinds can block unwanted solar heat in summer. In winter, low-e films, multiple glazings, and movable insulation will diminish heat losses. Look for skylights that use insulated glass, that are openable, and that are sold with options such as reflector shades, blinds, screens, low-e films, and sun-control films. Most importantly, look for units with tight seals and weather stripping, or you will have cold-air infiltration in winter.

Energy-Conserving Windows

Heat Mirror 88, 77, 55
Southwall Technologies
1029 Corporation Way
Palo Alto, CA 94303
(Model 88 is double glazed with low-e
film suspended between glazings, R-4.3;
transmits 53 percent of solar energy,
71 percent of visible light. Models 77 and
55 are for sun control; transmit 46 and
28 percent of solar energy and 68 and
49 percent of visible light, respectively.)

Sungate 100
PPG Industries
One PPG Pl.
Pittsburgh, PA 15272
(Double-glazed, low-e insulated glass,
R-2.9; transmits 55 percent of solar
energy, 73 percent of visible light.)

3M Sungain Window Film
3M/Energy Control Products
3M Center
St. Paul, MN 55144-1000
(Antireflective film sandwiched in triple
and quadruple units offering R-2.8 and
R-3.8, respectively. Film transmits
63 percent of solar energy, 98 percent of
visible light.)

Skylights

Skymaster
Tub-Master Corp.
413 Virginia Dr.
Orlando, FL 32803
(Domed and flat plastic; double-glazed
units.)

Sunrise, Galaxy Roof Windows
Roto Frank of America, Inc.
Research Park
P.O. Box 599
Chester, CT 06412

(Double-glazed, low-e insulated glass;
screens, blinds, and other options
available.)

Velux Skylight, Roof Window
Velux-America, Inc.
P.O. Box 3268
Greenwood, SC 29646
(Double-glazed, insulated glass units.)

Movable Window Insulation

Comfort Shade
Dirt Road Co.
R.D. 1, Box 260
Waitsfield, VT 05673
(Mylar and fiberglass, R-6.7; skylight and
greenhouse shades also available.)

Energy Shield
Energy Saving Marketplace
82 Boston Post Rd.
Waterford, CT 06385
(Fiberfill and Mylar quilt, R-4.3.)

Insul Louver, Insul Shutter
First Low Products, Inc.
P.O. Box 888
69 Island St.
Keene, NH 03431
(Foil-faced, rigid polyisocyanurate core in
hardwood veneer, R-8.)

Sun Blind
FTR Insulated Shutter System
5725 Arapahoe
Boulder, CO 80303
(Thermax core, R-9.)

Window Quilt, Showcase
Appropriate Technology Corp.
P.O. Box 975
Brattleboro, VT 05301
(Window Quilt is Mylar and fiberfill with
cotton cover, R-4.3. Showcase is
insulating roll-down shade, R-6.)

Movable Window Sun Control

Canvas Pacifica Castec, Inc.
7531 Coldwater Canyon Ave.
North Hollywood, CA 91605
(Natural canvas; shading coefficient of .33 to .35.)

Galaxy Sun Controller
Levolar Lorentzen, Inc.
1280 Wall Street W
Lyndhurst, NJ 07071
(Aluminum venetian blind; shading coefficient of .38 to .88; other models include one-way see-through and high-reflection options.)

Haluscreen
Haluscreen
2664B Mercantile Dr.
Rancho Cordove, CA 95670
(PVC-coated fiberglass sun screen; shading coefficient for exterior unit of .15 to .21.)

Insulated Solarium Shading System
Thermal Designs, Inc.
921 Walnut
Boulder, CO 80302
(Cotton canvas with Mylar backing.)

Sun Screen
Phifer Wire Products
P.O. Box 1700
Tuscaloosa, AL 35403
(Woven fiberglass solar screening; shading coefficient of .23 to .42.)

Interior Storm Windows

3M Window Insulation Kits
3M/Energy Control Products
3M Center
St. Paul, MN 55144-1000
(Nonrigid clear plastic film with tape edge seals. Also available: rigid acrylic glazing with magnetic seal.)

Window Films

Scotchtint Y-2728
3M/Energy Control Products
3M Center
St. Paul, MN 55144-1000
(Transmits 26 percent of solar energy, 18 percent of visible light; reflective silver.)

Sun Guard DRLW150 GG25
Metallized Products
2544 Terminal Dr. S
St. Petersburg, FL 33712
(Transmits 45 percent of solar energy, 27 percent of visible light; gray tint.)

Winter Gold-80
Deposition Technology, Inc. 7670
8953 Carrol Way
San Diego, CA 97121
(R-2.6 on double glazing, transmits 41 percent of solar energy, 53 percent of visible light; gold tint.)

Section 5
Timing Devices

Energy-conserving timing devices allow you to pattern the operation of heating and cooling equipment, lights, and other electrical appliances to coincide more appropriately with household energy cycles and human activities. Substantial fuel savings can be derived from thermostat setbacks on heating equipment and the adoption of daily off-cycles for water heaters. The cost to operate air conditioners can be similarly reduced with timers. For people charged under off-peak/on-peak rate schedules, timers can hold their bills down to minimal levels.

Until recently, windup mechanical units similar to kitchen timers having the ability to preset a single temperature or on/off option were widely used to control residential electrical appliances. A somewhat more sophisticated device, the clock thermostat, allows more time-period and temperature presetting options during a 24-hour period. It is designed to control the operation of furnaces and air conditioners. Both mechanical and clock thermostats can save energy.

A reduction of 1 to 2 percent of annual heating costs can be expected for each 8-hour period during which you have a 1°F setback. Thus, an 8-hour nighttime setback of 10°F should save you at least 10 percent. A 15°F night setback (e.g., 70°F to 55°F) plus a 3°F day setback for 8 hours should bring an 18 to 20 percent reduction to your heating bill. Air conditioning costs will follow an analogous drop if timers are used to turn up the temperature at various times. Other electrical equipment such as lights and dehumidifiers can also benefit from timers.

Recently, microelectronic thermostat-control devices have been introduced and may revolutionize the way we use residential heating and cooling systems. These devices can be preprogrammed to control the operation of energy-using equipment for an entire week, and many daily options are possible. For example, you can establish a pattern of up to six temperature settings and six time periods per day for heating or cooling.

Most of these units include the capability to establish a five-day pattern for weekdays and a separate two-day weekend pattern as well. They anticipate how long in advance a furnace must fire up in order to meet your preset temperature and time scenarios. To help boost energy efficiency, options include longer on and off cycles for furnaces. Rational adjustments to heating and cooling cycles that would be difficult to carry out otherwise are conveniently implemented with microelectronic control systems.

Microelectronic control over appliances and lighting is also possible. Clothes washers, dryers, security systems, ovens, and Crockpots can all be directed from the same control box, which is plugged into any household outlet. Control signals are transmitted through house wiring to individual plug-in modules that are designed to be used with either lights or appliances. Hot water systems or any appliance using 220-volt power would require a separate compatible control system.

Space heating, water heating, and cooling system setbacks provide economic benefits. Redistributing the times of operation of other appliances to fit in with differing electric rates during different times of the day can also cut costs. In some locations, electric company customers pay according to the highest amount of electrical power they draw for any 15-minute period during the month. Control systems can be programmed to balance out the use of large appliances in order to reduce monthly costs. Some electronic control units will monitor power consumption and cut off one or two appliances for a brief period if consumption rises too high.

Electronic control systems are easy to install—a serviceperson can do the job in a few minutes. Clock thermostats are normally substituted for existing wall thermostats and are connected to the two wires that come through the wall behind the thermostat.

Timing Devices

All Seasons
Quad Six, Inc.
3753 Plaza Dr.
Ann Arbor, MI 48104
(Preprogrammed; seven setback periods per day; adjusts to weather conditions; cycling for air conditioning.)

Electric Water Heater Time Switch
Solar Components Corp.
P.O. Box 237
Manchester, NH 03105
(Up to 12 on/off operations per day; 250 volts.)

Honeywell Clock Thermostat Series
Honeywell, Inc.
Honeywell Telemarketing Center
P.O. Box 524,
Honeywell Plaza
Minneapolis, MN 55408
(Numerous units with up to seven different daily programs; adjusts to weather conditions.)

Magic-Stat Furnace Thermostat
Solar Components Corp.
P.O. Box 237
Manchester, NH 03105
(Six temperature settings with up to seven different daily programs; adjusts to weather conditions.)

Programmable Command Console Outfit
Sears, Roebuck and Co.
Sears Tower
Chicago, IL 60684
(Turns up to eight lights or appliances on/off up to twice a day each.)

Tightwatt
Trane Dealer Products Group
Troup Highway
Tyler, TX 75711
(Used with heat pumps; two setbacks per day.)

Section 6
Refrigerators and Freezers

Refrigerators have evolved to such an extent that they are one of the most energy-efficient appliances available. In 1972, the average new refrigerator used 1,700 kwh a year. Today, the average new model uses about 1,200 kwh a year, and the most efficient models consume only 850 to 1,000 kwh a year.

The refrigerator works by using a compressor, a refrigerant-filled condenser coil, and a fan. Together, these components extract heat from the interior of the refrigerator. Higher efficiency comes from increasing the insulation in the refrigerator's surfaces (to hold cold in and heat out) and from improving the door gaskets (in order to reduce warm-air infiltration). Improved compressors and fan motors also boost efficiency.

The freezer compartment should be insulated from the refrigeration compartment. This reduces freezer energy loss when the refrigerator door is opened and therefore increases the time period between automatic defrost cycles from hours to days.

To reduce moisture buildup during periods of high humidity, most refrigerators have a small electric heater operating inside the door. (Yes, believe it or not, most refrigerators have built-in heaters!) Newer models have a "power saver" switch so that you can cut power to these electric coils in nonhumid weather.

Freezers with high efficiency ratings have similar features. Chest freezers come on less often than uprights because they are opened from the top, so they do not dump their cold air when they are opened. Good seals and thick insulation on a 16-cubic-foot model can mean an annual power consumption of less than 800 kwh.

Manual-Defrost Refrigerators

Kenmore Model 65111N Refrigerator/ Freezer
Kenmore Model 65011N Refrigerator/ Freezer
Sears, Roebuck and Co.
Sears Tower
Chicago, IL 60684

(Model 65111N: 11.6 cu. ft.; manual-defrost freezer and refrigerator; less than 500 kwh annual power use. Model 65011N: 10.6 cu. ft.; manual-defrost freezer, self-defrost refrigerator; approx. 700 kwh annual power use.)

Self-Defrosting Refrigerator/Freezers

Amana T Series
Amana Refrigeration, Inc.
Amana, IA 52204
(14.2 to 19.9 cu. ft. models; 865 to 1,020 kwh annual power use.)

Kenmore High-Efficiency Models
Sears, Roebuck and Co.
Sears Tower
Chicago, IL 60684
(17.7, 18, and 19.6 cu. ft. models; 978 to 1,156 kwh annual power use.)

Whirlpool ET Series
Whirlpool Corp.
Administrative Center
2000 N. U.S. 33
Benton Harbor, MI 49022
(17.2 to 19.5 cu. ft. models; 880 to 1,090 kwh annual power use.)

Freezers

Kenmore Upright Models, Chest Model
Sears, Roebuck and Co.
Sears Tower
Chicago, IL 60684
(Upright models: 13.1 to 20 cu. ft.; manual defrost; 730 to 810 kwh annual power use. Chest model: 15.1 cu. ft.; manual defrost; 710 kwh annual power use.)

Whirlpool K and L Series
Whirlpool Corp.
Administrative Center
2000 N. U.S. 33
Benton Harbor, MI 49022
(Upright models: 15.1 to 20 cu. ft.; manual defrost; 770 to 810 kwh annual power use. Upright models: 15.2 to 19.6 cu. ft.; auto-defrost; 1,010 to 1,245 kwh annual power use.)

Woods Models E420, E581
W. C. Wood Co.
P.O. Box 750
5 Arthur St. S
Guelph, ON N1H 6L9
Canada
(Chest models: 14.9 to 17.6 cu. ft.; manual defrost; 708 to 756 kwh annual power use.)

Section 7

Ranges and Ovens

New electric ranges tend to be more energy efficient than older models. But the savings will probably be quite small, and the cost of buying a new range can be exorbitant. In general, the best approach is to switch to gas cooking.

Gas ranges cost about half as much to operate as electric ranges. Get a gas model with a pilotless ignition—this can save up to a third of your outlay for fuel. Also seek out a smaller oven, preferably one that has insulation in its walls. Both features will save energy. A tight-fitting windowless door can also contribute to energy efficiency.

If you are determined to cook electrically, consider a convection oven. Electric convection ovens can cut your electric energy use in half.

Gas Ranges and Ovens

Kenmore Model 73931N Gas Range
Sears, Roebuck and Co.
Sears Tower
Chicago, IL 60684
(Natural gas or liquid propane; electric ignition.)

Tappan Model 302603 Gas Range
O'Keefe and Merritt Model 308482
Tappan, O'Keefe and Merritt Co.
222 Chambers Rd.
P.O. Box 606
Mansfield, OH 44901
(Natural gas or liquid propane; electric ignition.)

Wards Model 2472 Gas Range
Montgomery Ward and Co.
1000 S. Monroe St.
Baltimore, MD 21232
(Natural gas or liquid propane; electric ignition.)

Electric Ranges and Ovens

Farberware 460/5
Farberware
1500 Bassett Ave.
Bronx, NY 10461
(Convection oven.)

Fasar Magnetic Induction Cooktop
Bacun, Inc.
2801 Burton St.
Burbank, CA 91504
(Glass-top induction unit.)

Section 8
Energy-Efficient Air Conditioners

The most effective way to cut air-conditioner power consumption is to counteract the sources of excess heat with rafter foil, attic insulation, sun screens, and sun-blocking devices over windows. However, despite the various possibilities for low-energy and no-energy techniques for cooling a building, air conditioners will continue to be the most popular cooling technology in America for some time to come.

Major energy savings can accrue through conversion to the most energy-efficient air-conditioning equipment. Central and room air conditioners are rated according to their Btu capacity per hour. Central units are rated at 20,000 to

80,000 Btu per hour (i.e., they can remove this much heat in an hour), and room-size models range from 5,000 to 20,000 Btu per hour. The Btu capacity you need in order to meet your home's cooling load will vary according to how well insulated and tight your home is. Generally speaking, a well-insulated house will lose 15 Btu or less per hour per square foot of floor area. Thus, a 1,800-square-foot house would need an air conditioner with a capacity of around 27,000 Btu per hour. The same size house would lose about 40 Btu per hour if it were uninsulated, so then it would require a system with a capacity of 72,000 Btu per hour.

A 15-by-20-foot room in a well-insulated house would require a room unit with a capacity of around 4,500 Btu per hour, whereas the same size room in an uninsulated dwelling would require a 12,000-Btu-per-hour air conditioner. The advantage of "tightening up" the house is therefore obvious.

You want to get as many Btu per hour as you possibly can for every kwh you use. The Energy Efficiency Rating (EER) of all air conditioning equipment is listed on new models or can be obtained from your local dealer for your current model. The higher this figure is—the range is from 1 to around 14—the more Btu per hour you will get for every kwh you use. If air conditioning currently costs you $800 a year and your present equipment has an EER of 6, a new model with a EER of 11 will bring your bill down to about $440.

Boosting the efficiency of your current model is possible. Since an air conditioner works harder when it is in a hot environment, shading it will help to make its job easier and thus save you energy. Using a misting device to spray your air conditioner's condenser coils with cool water will facilitate evaporative cooling and make it easier for the condenser to shed heat. The use of a clock thermostat is also of major importance.

Central air conditioners have an outside compressor, coil, and fan. These are linked to your home via a coil mounted inside the same ductwork used for winter forced-air heat distribution. Household air is circulated through the duct system and it loses some of its heat and moisture to a refrigerant gas—usually freon—circulating within the coil. As the refrigerant in the coil passes through the compressor, it gives up much of its heat and moisture. The better the equipment is at facilitating this process, the higher its efficiency. Newer cooling systems use smaller but better compressors, a larger outside coil, and a stronger fan to bring down the condenser's temperature. Another major change in air conditioner technology is the use of improved control systems that do a better job of monitoring refrigerant pressure and temperature as well as both outdoor and room temperatures. This system of sensors is in turn used to fine-tune the operation of the compressor.

Room air conditioners operate in a similar way except that the entire system is much more compact and is confined to a single location, usually in a window. Because room units do not physically separate the indoor and outdoor components, EER ratings run a bit lower than with central air conditioners. On the other hand, they allow zoning (i.e., selective cooling of a few rooms) and they require no added ductwork installation. Reverse-cycle dual-function models will also provide heat in the cooler months. Look for a room air conditioner with an EER of 9 to 10, and do not buy a larger unit than you need based on the area to be cooled and how well insulated your house is.

Central Air Conditioners

Premium Round One
Carrier Corp.
P.O. Box 4800, Carrier Pkwy.
Syracuse, NY 13221
(16,200- to 58,000-Btu capacity;
EER 11.6.)

Rheem RAGC Series
Rheem Manufacturing Co.
5780 Peachtree, Suite 400
Dunwoody Rd.
Atlanta, GA 30342
(23,600- to 36,000-Btu capacity;
EER 12.1 to 12.3.)

Room Air Conditioners

Carrier 51 Series
Carrier Corp.
P.O. Box 4800, Carrier Pkwy.
Syracuse, NY 13221
(5,400- to 12,050-Btu capacity;
EER 9 to 11.)

Friedrich SP, SS, SM, SA Series
Friedrich Air Conditioning and
Refrigeration Co.
P.O. Box 1540
San Antonio, TX 78295
(6,500- to 13,600-Btu capacity; EER 9 to
11.5.)

Whirlpool AC 1402XMO
Whirlpool Corp.
Administrative Center
2000 N. U.S. 33
Benton Harbor, MI 49022
(14,000-Btu capacity; EER 9.9.)

Misting Devices

Automatic Mister
Thermodyne, Inc., ACES Div.
2675 Iliff St.
Boulder, CO 80303
(Automatic solid-state system; connects
to plumbing; .5 gal. water use per hr.)

Cool Mist Air Conditioner Companion
Cool Mist, Inc.
4801 N. State, Suite 406
Jackson, MS 39206
(Fully automatic; nonelectric; air-
discharge system.)

Section 9
Fans

Fans are simple, low-energy devices that can cut up to 50 percent from air conditioning bills. They can also help to reduce heating costs. Fans serve three purposes: (1) they are used to redistribute either warm or cool air within a house; (2) they facilitate the removal of hot air from a room or an entire house in summer to make way for cooler outside air; (3) they provide evaporative cooling to the human body.

It is important to know what type of fan is most appropriate for your particular objective (e.g., ventilating your house or creating a gentle breeze to cool you). You should also decide, based on your specific purpose, how much air you need to move over a given time period. This will depend on the size of the area you wish to affect.

The size of the fan—the area the blade sweeps—and its rotational speed measured in revolutions per minute (RPM) will influence the amount of air it can move in cubic feet per minute (cfm). If you want to vent all the air in a 10-by-12-foot room having a 7-foot ceiling, and if you want to do it in 1 minute, the fan would need to move approximately 840 cfm. If you needed to empty the room only once every 4 minutes, you would need a fan with a 210-cfm rating.

You should also be concerned with the efficiency of the fan. If you divide the fan's cfm rating by its wattage, you can use this figure to make comparisons. Thus, a fan that uses 50 watts to move 200 cfm of air gives you 4 cubic feet per watt. It is therefore as efficient as a fan that uses 25 watts to move 100 cfm or one that uses 100 watts to move 400 cfm.

Let's look at each of the three main uses for fans.

1 Improving the circulation of warm air within homes has become a concern in this age of wood and coal stoves, minifurnaces, kerosene heaters, and solar space heating. Heat flow between rooms and between greenhouses and the house's main living areas can be improved with the use of small fans mounted within walls. Several companies make fans that are designed to fit between wall studs. In many cases, these fans can be controlled by thermostats.

Since the purpose of interior-wall fans is to move air without creating a discomforting draft, blade rotation tends to be slow, so the amount of air moved is around 150 to 200 cfm. However, a fan used to vent a solar greenhouse should move air quickly enough to prevent the greenhouse from overheating. A 10-foot-long by 8-foot-wide greenhouse with an average height of 8 feet has a volume of 640 cubic feet. On sunny days at noon, you will want to move air very quickly from the greenhouse into the house, since this air will contain a substantial amount of heat. On partly cloudy days or in the early or later part of the day when there is less sunlight, you will not need to exhaust the greenhouse as quickly.

You can control air flow by using a variable-speed fan or a thermostatically controlled fan. Varying fan speeds manually according to greenhouse temperatures can be laborious; for this reason, I prefer thermostatic control. The fan would automatically work longer on hot, sunny days than on dim days. Note that a 640-cubic-foot greenhouse venting 95°F air for 5 hours into your home's interior at one complete air change per minute could raise the temperature of a well-insulated 1,200-cubic-foot area from 55°F to about 75°F.

The wall fan is ideal for distributing cool air between rooms in summer. People frequently purchase room air conditioners that have a greater cooling capacity than is needed for the space being cooled. If you are in this situation, you could use a wall fan to send excess cool air into an adjoining room. Although some people argue that the fan should be installed near the floor, where the air is coolest, a fan mounted closer to the ceiling brings cool air into the upper section of the next room and thus allows for better mixing. Installing the fan at this higher position also makes the fan ideal for winter heat distribution.

Another concern with redistribution of heat relates to heat stratification between the ceiling and floor. Heat rises and then tends to stay in the upper reaches of a room. This can be particularly troublesome if you have very high ceilings. Small, low-RPM ceiling fans can retrieve this heated air and send it back down to where you need it. The fans produce better mixing of air currents and a more uniform temperature at all levels in the room. Most such fans use around 20 watts and, according to manufacturers, they can destratify up to 500 square feet, saving up to 15 percent annually on heating bills.

Ceiling-mounted "paddle" fans have been used as cooling devices for years. Their slow-moving blades can also do a reasonable job of mixing air in a room and hence they counteract heat stratification in winter. Unfortunately, the breeze created often causes people below to feel cooler in winter, so although temperatures may be more balanced within a room, you may be less comfortable than before.

2 Ventilation is the primary approach used to reduce the need for air conditioning. The idea is to get hot air out of a room or house and to get cooler outside air into it. This requires that you have some awareness of the relationships between outside and inside air temperatures so that you do not bring into your house air that is hotter than the air you already have inside. Generally, this means you should not ventilate before 4:00 or 5:00 P.M. unless you can draw outside air from a cool forested area or from over water. Three fans commonly used for venting hot interior air are the attic, window, and whole-house models.

The attic fan is installed in a gable vent in order to expel heated air from the attic. Although it does a fair job of reducing attic temperatures, the use of rafter foil and attic insulation is a better way to combat heat attacks from above.

Window fans can be powerful enough to clear a huge volume of hot air out of your house in a short time. Quality units can operate in the 3000- to 5000-cfm range and will provide the equivalent of over 20 whole-house air changes in an hour. It helps to have a house with as few horizontal obstructions to air flow as possible, and the fan should be mounted in a high window. The major advantage of window fans is their ease of installation.

Whole-house fans are permanently mounted in the attic. They draw air from inside the house and vent it into the attic where it exits through gable vents (you may need to increase the size of the vents to facilitate this process). The fan you want will exhaust an amount of air equal to the volume of your house once every minute. Thus, a 1,500-square-foot house with 7-foot ceilings will need a 10,000-cfm fan. The fan will not only get rid of hot air but will also cool down hot surfaces within your home, and the breeze it creates will cool you. Keep in mind, however, that a whole-house fan creates a very large hole in your ceiling. You must carefully seal it off in winter or it could cause serious heat losses.

3 The use of portable box fans, oscillating fans, and ceiling-mounted paddle fans to create a cooling air flow over your skin should not be dismissed as old-fashioned. It is an effective, low-cost way to maintain summertime comfort. Unless humidity is high, you should be able to avoid the need to air condition even with temperatures in the mid-80s by using these fans.

Fans

Heat Mover
Hutch Manufacturing Co.
Energy Products Div.
Box 350
200 Commerce St.
Loudon, TN 37774
(Wall-mounted fans for rooms or sunspaces; 160 cfm; 75 watts.)

Stratojet
Rust Industries, Inc.
3504 S. Pennsylvania St.
Englewood, CO 80110
(Destratifier for 8-foot to 20-foot ceilings; 40 cfm at 10 watts to 150 cfm at 58 watts.)

Window and Whole-House Models
Sears, Roebuck and Co.
Sears Tower
Chicago, IL 60684
(A number of fans available through the catalog.)

Section 10
Air-to-Air Heat Pumps

A baseboard electric-resistance heating system generates heat directly from electricity whereas the air-to-air heat pump uses electricity to run a compressor that extracts heat from surrounding air. Because the heat pump only has to capture heat and not create it, it can provide more energy to your living space than it uses to run the compressor. Resistance systems can only provide as much heat as electricity contains—about 3,412 Btu per kwh. Thus, the conversion from electric-resistance heating to a heat pump can save you money.

The heat pump is essentially an air conditioner operating in reverse. In fact, many units sold are "reverse-cycle models" that offer both heating and cooling. The heat pump contains a coil that runs in a loop between the inside of your house and the outside. The coil contains freon that circulates through the coil. Just before the freon goes outside, it passes through an expansion valve. A characteristic of freon is that when it expands (i.e., when it is under less pressure), it changes from a liquid to a gas.

When the gaseous freon passes through the outdoor section of the coil, it absorbs heat from the surrounding air. (Even when the air feels cool to us, it contains a substantial amount of heat energy—45°F may seem cold to human bodies, for example, but it is very much warmer than absolute zero, -460°F, the temperature at which there is absolutely no heat.) The freon then reenters the house as a warm gas. The compressor increases the pressure on the freon, which becomes liquid again, a process that raises the freon's temperature further. The liquid freon is hotter than the air inside the home, so it releases its warmth to the interior environment. A fan boosts the efficiency of the heat-exchange process by blowing air across the condenser coil where the exchange occurs.

The warmer it is outside, the more heat energy is absorbed per kilowatt of power used to run the compressor and the higher the operating efficiency of the heat pump. The Coefficient of Performance (COP) indicates the ratio of how many Btu of energy we extract from the environment compared to the amount of energy used to run the equipment. Higher outdoor temperatures will lead a heat pump to produce more Btu of heat for each kwh used, thus raising the system's efficiency, while lower temperatures will mean a lower COP. The most efficient heat pumps can provide over three times more Btu to your house than you burn up in electricity (i.e., a COP of 3 for outside temperatures above 50°F) and twice the Btu you use at lower temperatures (i.e., a COP of 2 at around 32°F).

The fact that efficiencies fall as outdoor temperatures fall creates a problem. There is heat available in the outside air, but the freon becomes very dense and the compressor has to work harder to extract heat. Therefore, at a predetermined temperature setting, an electrical-resistance backup heater inside the heat pump comes on to supplement the heat produced by the freon coil. This temperature setting varies according to the design of the heat pump.

A related problem in cold weather is moisture buildup around the outdoor section of the freon coil—this moisture turns to ice at low temperatures, further impeding the heat pump's performance. New heat pumps use an electrical defroster to melt the ice, but this means greater energy consumption. The far greater popularity of heat pumps in regions of the country where winter temperatures are mild and air conditioning is needed in summer reflects the problems of reduced efficiency and system icing associated with cold-climate operation.

Note that add-on heat pumps are available for use with your existing oil or gas furnace. The above problems are overcome to a degree with this two-mode system in that the pump operates in sequence with your existing system. Thus, the heat pump runs during periods of higher temperatures (when it is at peak efficiency) and the furnace operates at lower outside temperatures (when it runs longer and more frequently, and thus has lower flue losses). Add-on units will also operate as air conditioners in summer. Air conditioning equipment efficiency is measured by EER—you want a system with a high EER as well as a high COP. (For more information about air conditioning, see section 8, "Energy-Efficient Air Conditioners.")

Remember that the heat pump is an electrical appliance and the higher your local electric rates, the less reasonable it is to select any electrical heating system. If you do purchase a heat pump, look for a unit with efficient and reliable compressors and fan motors, a corrosion-resistant coil, an automatic defrost system for the outdoor coil, a crankcase heater to increase compressor longevity, and timing controls that will prevent rapid on/off cycling.

Heat Pumps

Friedrich Reverse Cycle Heat Pumps
Friedrich Co.
P.O. Box 1540
San Antonio, TX 78295
(EER 9 to 11; COP 2.4 to 2.8; 5,000- to 20,000-Btu room-size unit.)

Lennox HP16211V
Lennox Industries, Inc.
P.O. Box 809000
Dallas, TX 75380
(EER 8.5 to 9.4; COP 2.65 to 2.9; 24,000- to 60,000-Btu central unit.)

Section 11
Energy-Efficient Furnaces

Despite the various alternative space-heating systems that have emerged in response to rising energy costs, oil and gas furnaces continue to be the most popular choices for home heating. A minority of households have switched to wood, solar, or coal, but high heating costs have been attacked primarily through the use of insulation, caulk, and weather stripping to make houses tighter, and by modifying existing heating systems to boost their overall efficiencies. Since electric-resistance heaters already operate at maximum efficiency, there is little room for improvement of such systems. The use of the heat pump as a substitute electric system has some economic merit, but not as much as the installation of an efficient oil or gas furnace.

The measure of furnace efficiency is its Annual Fuel Utilization Efficiency (AFUE), which indicates the percentage of potential energy in the fuel that gets converted to usable heat over a year of operation. Unmodified older gas furnaces have an AFUE of about 55 percent, while oil units of the same vintage are about 60 percent efficient.

Techniques used to boost AFUE values on both existing and new furnaces generally involve improvement of the combustion process and extracting a greater proportion of the heat that is generated. The latter is done through improved heat-exchange techniques and by curtailing flue losses.

Installing smaller fuel nozzles in oil furnaces, for example, will slow down the rate of fuel consumption so that combustion occurs over longer periods of time without using more fuel. Furnaces cycle on and off, generating heat and then sitting idle until the house needs more heat. During the idle periods, heat from the furnace escapes up the flue. The use of smaller fuel nozzles reduces the length of the idle periods, which in turn reduces heat losses up the flue. This procedure, known as "derating," can increase oil-burner efficiencies by 5 to 10 percent.

Gas furnaces that use an electronic ignition instead of a continuously burning pilot light will be about 5 percent more efficient. Flue dampers that automatically close off the flue pipe when the burner is not on can save 10 percent on fuel costs for oil, and 5 to 8 percent with gas systems.

Inefficient older furnaces are often retrofitted with flue heat-recovery equipment. In some instances where flue temperatures are less than 600°F to begin with, these add-on devices can bring flue temperatures down so far that water condensation from exiting gases will occur. Thus, extracting heat from an old

furnace's flue will boost energy efficiency (the heat taken from the flue is used to heat your house), but it might also lead to corrosion of metal surfaces and reduce the life expectancy of the furnace. Newer furnaces that extract flue heat are designed to deal with condensation and will give you over 80 percent AFUE.

New, "noncondensing" furnaces incorporate energy-conserving features but provide no more than 70 to 80 percent efficiency. To reach higher AFUE levels, you must purchase a "condensing" furnace. Normally about 10 percent of the heat generated through combustion is contained within water vapor that goes up your chimney. Condensing furnaces capture this heat via large heat-exchange units. These reduce flue gas temperatures to below 200°F. Condensing furnaces use stainless-steel heat exchangers that will not corrode. In superefficient furnaces, a second and sometimes a third heat exchanger is used. So much of the available heat is extracted that flue gases drop to as low as 100°F and the flue can be made of inexpensive PVC pipe.

The most efficient oil furnaces reach 85 to 88 percent AFUE, while gas equipment can achieve 90 to 97 percent AFUE. When used in well-insulated houses, these systems will bring the cost of heating with oil down to 42 percent of the cost of electric heat, and they bring the cost of heating with gas down to no more than 30 percent of the cost of electric heat.

High-efficiency furnaces range in output from about 40,000 to 160,000 Btu per hour. Well-insulated homes need only 10,000 to 40,000 Btu per hour to maintain comfort. Oversizing a furnace by more than 25 percent can lead to major inefficiencies in its operation. Thus we face a dilemma in that superefficient furnaces have high outputs and are designed to be used in dwellings that are not all that well insulated. If your house is tight, look for a smaller model and make sure it has low off-cycle heat losses (i.e., make sure it doesn't lose a lot of heat during idle periods), since it will be off much of the time.

Minifurnaces present an alternative technology particularly suited to well-insulated homes. These furnaces have outputs of 10,000 to 30,000 Btu per hour. They are normally sought as supplemental heaters, but in tight homes they often can meet the total annual load, although some mechanism for heat distribution such as several low-energy fans will probably be necessary.

If your home is tight enough that a minifurnace can meet your heating load, your level of fresh-air ventilation is probably not high (the house is tight, so stale indoor air does not escape readily). In this situation, you do not want to create indoor air pollution by installing a nonventing minifurnace. On the other hand, vented systems (those using flues) rarely exceed 60 percent efficiency and can be expensive to install. Thus, what you want is a direct-venting or sealed-combustion unit. This type of minifurnace does not contribute to indoor air pollution, since it uses outside air for combustion and exhausts the combustion by-products through a vent that is easily installed on an exterior wall. Direct-venting minifurnaces should give you around 90 percent efficiency, and gas equipment will be your best bargain.

Furnaces

Amana Energy Command
Amana Refrigeration, Inc.
Amana, IA 52204
(76,000- to 95,000-Btu output;
95 percent AFUE; gas.)

Carrier 58SS Super Saver
Carrier 58SX Weathermaker
United Technologies, Carrier Corp.
P.O. Box 4800, Carrier Pkwy.
Syracuse, NY 13221
(58SS Super Saver: 39,000- to 139,000-
Btu output; 83 percent AFUE; induced
draft combustion; noncondensing; gas.
58SX Weathermaker: 61,000- to
102,000-Btu output; 91 percent AFUE;
double heat exchangers; condensing unit;
gas.)

Duo-Matic/Olsen Ultramax
Duo-Matic/Olsen, Inc.
56 S. Squirrel Rd.
Pontiac, MI 48057
(90,000- to 145,000-Btu output;
88 percent AFUE; oil.)

Empire Gemini II, Mini-Line Series,
Upright Counterflow
Empire Stove Co.
P.O. Box 529
918 Freeburg Ave.
Belleville, IL 62222
(Gemini II: 15,000- to 30,000-Btu output;
83 percent AFUE; vented; gas. Mini-Line:
10,000- to 15,000-Btu output; 85 percent
AFUE; direct-venting; gas. Upright
Counterflow: 60,000-Btu output; direct-
venting; wall mounted; gas.)

Lennox Pulse
Lennox Industries, Inc.
P.O. Box 809000
Dallas, TX 75380-9000
(38,000- to 74,000-Btu output; 93 to
97 percent AFUE; gas.)

Whirlpool Tightfist II
Whirlpool Corp.
Administrative Center
2000 N. U.S. 33
Benton Harbor, MI 49022
(47,000- to 94,000-Btu output; 93 to
96 percent AFUE; gas.)

Minifurnaces

Cozy Direct-Vent Minifurnace
Louisville Tin and Stove Co.
Box 1079
S. 13th St.
Louisville, KY 40201
(Through-the-wall unit; 10,500-Btu
output; gas.)

Heil-Quaker Minifurnace Series
Heil-Quaker Corp.
1136 Heil-Quaker Blvd.
P.O. Box 3005
Laverne, TN 73076-1985
(Several flue-vented natural gas models
available.)

Kero-Sun Direct Vent Minifurnaces
Kero-Sun, Inc.
P.O. Box 549
Kent, CT 06757
(Through-the-wall or window-vented units;
19,000- to 32,000-Btu models;
88 percent efficiency; remote fuel tank
hookup advised.)

Sears Gas and Oil Space Heater Series
Minifurnace
Sears, Roebuck and Co.
Sears Tower
Chicago, IL 60684
(Gas: 15,000- and 45,000-Btu models.
Oil: 50,000-Btu output. All require flue
and chimney venting. Direct-venting
20,000-Btu gas model also available.)

Section 12
Passive-Solar Space Heating

The major systems for residential passive-solar space heating are direct-gain systems, attached greenhouses, and TAPs (thermosiphoning air panel systems) and window heaters. Each has advantages and disadvantages, and all are capable of meeting some of your heating load. The proportion they actually meet depends on your local climate and "insolation" (how much sunlight you receive), how tight your house is, and the size and design characteristics of the passive-solar system you use.

Direct-gain solar heating is the process of heating the house by allowing sunlight to shine directly into it. This process requires that a glazed area (i.e., windows) be incorporated into the house's south-facing exterior wall. The size of the glazed area depends on the size of your home and how well insulated it is. An extremely well insulated house requires far less south-facing window area, since it retains heat so much better than a poorly insulated house.

The solar greenhouse and its relative, the sunspace, are designed to capture the sun's energy through an extensive glazing surface. The greenhouse is used for supplemental space heating of interior living space and as a microenvironment for growing plants. The sunspace is essentially the same thing, except that plants are not grown in it.

The TAP and the window heater use flat-plate collector technology to capture heat on the south wall of the house. The units are externally mounted. TAPs are linked to the house's interior by means of vents in the wall; window heaters send their heat into the house through a window aperture.

The net energy gain per square foot of glazing is similar in all three systems as long as effective night insulation is used in a direct-gain system (i.e., the glazing should be covered with insulating shades, shutters, etc., at night). Without night insulation, direct-gain systems provide only about half the net heat contribution of greenhouses and TAPs per square foot. This is because at night, heat can escape from the house through the glazing on the south wall—the heat loss through south glass during long winter nights is only slightly less than the daytime heat gain. Greenhouses and TAP systems are less prone to this problem—residual heat inside a greenhouse or a TAP may escape from the system at night, but you can close off the system from the house, so very little heat from the house itself escapes through the system.

Here's more detailed information about each system:

☐ A direct-gain system with a glazed surface of less than 10 percent of the floor area of the home can provide a reasonable supply of winter Btu to help reduce heating bills, if your home is well insulated. With larger glazing areas, more heat is gained, which can lead to overheating unless some medium such as masonry or water-filled containers is provided to soak up excess heat during the day and release it slowly at night.

In hot climates, direct-gain systems can involve a cooling penalty (i.e., sunlight entering the house makes cooling equipment work harder to maintain comfort levels). Overhangs and shading devices will not completely eliminate this unwanted solar heating in summer because some reflected and diffuse radiation will still pass through the glass.

2 Greenhouses and sunspaces are, in essence, externally mounted collectors, and as such they exhibit certain advantages over direct-gain systems. The design of greenhouses and the materials used in their construction are the primary factors determining how effective they will be.

Double-pane insulated glass is generally the best glazing for solar greenhouses. Fiberglass, acrylic, and polycarbonate glazings tend to be cheaper and lighter in weight, but they are prone to deterioration and have a shorter life expectancy. Glazings with heat-reflecting or sun-blocking coatings are also available.

A greenhouse's framing is usually constructed of wood or aluminum. Wood has a better insulation value, but it is adversely affected by the elements and requires maintenance. Redwood and cedar hold up best. Aluminum requires minimal upkeep but it has a very low R-value. Quality greenhouse and sunspace kits use a thermal break of low-conductive material to boost framing R-values.

Glazing angles are also important. South-facing glass that is tilted up to 30 degrees from vertical will give optimal solar heating during the winter. Glazed greenhouse roofs are difficult to insulate at night, and even during the day a glazed roof loses substantial heat through conduction. This is because heat rises naturally within the greenhouse, and the temperature can reach well over 100°F at the roof level. Since heat losses are largely determined by the difference between outside and inside temperatures, and losses increase as the gap between these temperatures widens, you can waste much of your valuable energy. Paradoxically, despite these heat losses, a glazed roof can also create significant greenhouse overheating during summer and even on some occasions in winter unless extensive heat-storage materials are provided. For these reasons, a non-glazed, well-insulated roof is usually best. This allows good winter heat gain through the glass on the south wall of the greenhouse and keeps things cooler during summer when the high sun's rays are strong on the roof. A single ventable skylight in the roof can be used for summer heat venting, or vents can be installed high on the east and west walls of the greenhouse.

The thermal performance and comfort of your greenhouse can be enhanced with the use of optional equipment. Insulating shades, shutters, etc., can boost the R-values of the glazed surfaces during sunless periods. Sun-control blinds or canvas is ideal for reducing solar heating when you do not want it. Wall fans will greatly improve the movement of heated air from a greenhouse into your living space, and exhaust fans can speed the venting of summer heat to the outside.

[3] TAPs and solar window box heaters were introduced as low-cost solar devices that can be built at home. A variety of mass-produced models are now on the market and generally tend to be more efficient than home-built models. They are reasonably priced and, as a retrofitting option, they present less of an installation hassle than direct-gain or greenhouse systems.

One option to consider is to buy a unit with a built-in fan and thermostatic controls. The use of an electric fan technically contradicts the definition of "passive" solar technology, since such technology is supposed to be free of mechanical devices. But low-energy fans do boost system efficiencies, and they are inexpensive to operate.

Solar Greenhouses and Sunspaces

Four Seasons System 4
Four Seasons Solar Products Corp.
425 Smith St.
Farmingdale, NY 11735
(Single- to triple-paned tempered glass; aluminum framing; shading and movable insulation; fans.)

Solar Additions Room
Solar Additions, Inc.
15 W. Main St.
Cambridge, NY 12816
(Double-paned tempered glass; redwood framing; operable windows; movable insulation; shading device; fan.)

Solar Room
Habitat
123 Elm St.
South Deerfield, MA 01373
(Double-paned tempered glass; cedar plus aluminum framing; shading device; walls and roof R-26.)

Sunspace
Solarium Systems, Inc.
209 E. 78th St.
Bloomington, MN 55420

(Double-paned tempered glass; pine plus aluminum framing; skylight; fans; movable insulation; shading device.)

TAPs and Window Heaters

E-100 Series Air Collectors
Environmental Energies, Inc.
P.O. Box 98, Front St.
Copemish, MI 49625
(Low-iron glass outer glazing; fluorocarbon film inner glazing; coated absorber plate.)

Solar Kal Airheater
Kalwall Corp., Solar Components Div.
P.O. Box 237
Manchester, NH 03105
(Double-glazed fiberglass polymer; aluminum absorber plate; 90-cfm fan; thermostat.)

Sun-Lite Window Box Heater
Kalwall Corp., Solar Components Div.
P.O. Box 237
Manchester, NH 03105
(Fiberglass polymer glazing; aluminum absorber plate; 100-cfm fan.)

Section 13
Wood Stoves

During the period of cheap fossil fuels from the 1940s through the early 1970s, heating homes with wood became relatively rare. Later, in the mid-1970s, wood use began to climb in response to oil price rises, even though many wood stoves of the time were quite inefficient. In the last ten years, however, the wood stove has evolved from a metal box in which uncontrolled combustion took place at 30 percent efficiency to become a finely tuned device, in some cases offering nearly 90 percent efficiency.

The overall efficiency of a wood stove depends on how completely the stove delivers to you the heat energy that is potentially available in the wood that you burn. Inefficient stoves let many of the Btu in the wood escape up the flue as gases and particulates, and they fail to adequately transfer heat from inside their fireboxes into the house's living space. Today's highly efficient stoves attack both these problems, extracting more heat from the wood and doing a better job of sending this heat into your rooms. Thus, wood—a renewable fuel—can now be burned at very respectable levels of efficiency, making it several times cheaper as a space-heating option than electricity.

New high-efficiency stoves often contain catalytic combustors that lower the temperature at which woodsmoke burns from above 1,000°F to about 500°F. This means that unburned gases and particulates that would go up the flue in a conventional stove are burned in a catalytic stove, giving off additional heat. In effect, the stove has two fires: a primary fire in the main combustion chamber, and a secondary fire where the catalytic combustor causes the smoke to burn. This also means there will be less air pollution and creosote buildup, which are both major problems with lower-efficiency stoves.

There are a number of catalytic stoves now on the market. Catalytic combustors can also be purchased for installation on an existing stove. These add-on combustors will not make an old stove as efficient as a stove specially designed for catalytic combustion, but they can boost an old stove's efficiency by 15 to 20 percent.

The cost of catalytic stoves can be $100 to $300 more than the cost of conventional airtight stoves. Add-on combustors run about $125.

The life expectancy of combustors is short, about one to five years. They are ceramic, woven wire, or metal units with a chemical coating that erodes over time. Replacement units cost about $100.

Another option is the purchase of one of several superefficient wood stoves

that are designed to get maximum heat without using a catalytic device. One model, for example, has a low-set primary combustion chamber and a separate secondary combustion chamber where a fresh stream of preheated air enters to provide additional oxygen, which in turn stimulates ignition of unburned gases. Heat is also forced to travel over baffles. This boosts transfer efficiency from the stove to the house by putting hot gases in contact with the stove surface for a longer time before they exit through the chimney.

Yet another option is the "forced draft" approach where fans or blowers channel fresh air over burning wood to intensify the heat given off. A stove using this approach may require some heat-storage medium such as brick to absorb the very substantial heat being generated. The storage medium will subsequently release this heat to the house after the fire has gone out.

An elaboration of this heat-storage process is the Russian stove. It is basically constructed as a massive masonry stove with a number of brick or cement-block baffles over which hot gases must travel before they exit the stove. The masonry stores up huge amounts of heat and, once warmed, it will continue to provide a steady output of heat to the living space over a very long period of time.

Catalytic Wood Stoves

American Golden Eagle
American Eagle Stoves
100 W. Drullard Ave.
Lancaster, NY 14086
(70 to 75 percent efficiency.)

Hearth Cat
Webster Stove
3112 La Salle St.
St. Louis, MO 63104
(65 to 70 percent efficiency.)

High-Efficiency Noncatalytic Stoves

Avenger D-1100
National Stove Works
Box 640
Cobleskill, NY 12043

Hearthstone 1,2,3
Hearthstone
Hearthstone Way
Morrisville, VT 05661

Jøtul 201 Turbo
Jøtul U.S.A., Inc.
343 Forest Ave.
P.O. Box 1157
Portland, ME 04104

Tempest Boiler
Dumont Industries
P.O. Box 148
Monmouth, ME 04259

TESS 148
Thermal Energy Storage System, Inc.
Box M, Mine Rd.
Kenvil, NJ 07847
(Heat storage masonry fireplace; 40 to 50 percent efficiency.)

Add-On Catalytic Combustors

Corning Catalytic Combustor
Sotz, Inc.
13676 N. Station Rd.
Columbia Station, OH 44028

Intensifire Catalytic Combustor
Catalytic Damper Corp.
P.O. Box 188
U.S. Rt. 522
Flint Hill, VA 22627

Nu-Tec Catalytic Combustor
Nu-Tec, Inc.
P.O. Box 908
East Greenwich, RI 02818

Smoke Consumer
Metal-Fab, Inc.
3025 May Ave.
P.O. Box 1138
Wichita, KS 67201

Section 14
Air-to-Air Heat Exchangers

The heat exchanger has been around for years, but it has been used mainly in industrial applications. Its purpose is to ventilate heated space without major heat loss. It does this by extracting the heat from air that it expels from the building and transferring this heat to fresh air that it draws in from the outside.

Residential units are relatively new in this country and have emerged in response to a growing need to ventilate tight, well-insulated houses. With air infiltration levels reduced to as low as only one air change every 2 hours, studies have shown significant increases in formaldehyde, smoke, radon, and other air pollutants indoors. An air-to-air heat exchanger fights this by removing stale air from the house with minimal heat loss. A unit able to remove 150 cubic feet of air per minute (cfm 150) would give a 1,500-square-foot house approximately one exchange of air per hour. There are also room-size models.

The technology itself can involve rotary exchangers or fixed exchangers used in conjunction with blowers. Room-size units use 25 to 60 watts per hour and cost about $10 to $20 a season to operate. Full-house units use 70 to 300 watts per hour and cost $25 to $100 a season to operate. Smaller units tend to have lower efficiencies, in the 60 to 70 percent range, whereas larger exchangers can reach efficiencies of 90 percent. The cost for house-size exchangers is generally between $300 and $800.

House-Size Heat Exchangers

Econofreshner
Berner International Corp.
P.O. Box 5205
New Castle, PA 16105
(30 to 60 cfm; 75 to 82 percent efficiency; rotary type exchanger.)

Lossnay VL-1500
Mitsubishi Electric Sales America, Inc.
3030 E. Victoria S
Compton, CA 90221
(30 to 70 cfm; 70 to 76 percent efficiency; fixed plate exchanger.)

Room-Size Heat Exchangers

AE-200
NuTone
Madison and Red Bank Rds.
Cincinnati, OH 45227
(150 to 200 cfm; 80 percent efficiency at high speed; rotary unit.)

Air Changer
Air Changer Co., Ltd.
334 Kind St. E, Suite 505
Toronto, ON M5A 1K8
Canada
(75 to 150 cfm; 76 percent efficiency at low speed; fixed plate.)

Z-Duct Series
Des Champs Laboratories, Inc.
P.O. Box 440
East Hanover, NJ 07936
(75 to 350 cfm; 85 percent efficiency at 150 cfm; fixed plate.)

Section 15
Dishwashers, Clothes Washers, and Dryers

The major energy cost of dishwashers and clothes washers results from their use of hot water rather than the power they consume for motor and pump operation. The conversion to gas water heating could cut your dishwashing and clothes washing bills by up to two-thirds.

Some dishwashers use 12 gallons of hot water per cycle while others use only 8 gallons. You want a model that uses less water. And make sure it has a switch that allows you to turn off the electric heating element. You will also save money with a model that has a water-temperature booster built in. Dishwashers normally require 140°F water temperatures, which is a higher setting than you should have on your water heater. Dishwashers with boosters are designed to heat water from 120°F (a good temperature for a water heater) to 140°F. Short-cycle dishwashers (i.e., those with brief wash cycles) also use less water, so look for this feature.

Clothes washers use very little power besides that related to hot water. Look for an adjustable water-level setting option. Some models allow you to reuse the same water for a second and third washing load.

Gas-powered clothes dryers cost only about a third as much to operate as electric models. Some new models can directly measure moisture inside the drum with an electric sensor. As soon as clothes are dry, the dryer will turn itself off.

Energy-Efficient Units

Kenmore Clothes Washer 26K33701N, Dryer 75921
Sears, Roebuck and Co.
Sears Tower
Chicago, IL 60684
(Washer has cold water rinse options; water-saver available. Dryer has moisture sensor; gas model costs 65 percent less to operate than electric one.)

Maytag A710 Clothes Washer
Maytag Co.
403 W. 4th St. N
Newton, IA 50208
(Four water-level options.)

Whirlpool Dishwasher DU7903XL, DU9903XL
Whirlpool Corp.
Administrative Center
2000 N. U.S. 33
Benton Harbor, MI 49022
(Low-energy wash cycle and air-dry options; DU9903XL has up to 9-hour delay setting, ideal for off-peak customers.)

Section 16
Alternative Generation of Electricity

This book focuses on how we might modify our consumption of electricity through conservation, fuel switching, and conversion to more efficient appliances. The production of electricity is a totally different issue, but it is relevant since, over the last decade or so, the technologies for home electricity production have become available to consumers. The options include solar photovoltaics, wind power, or small-scale hydroelectric power.

Unfortunately, the issue from the perspective of economics is quite simply stated: Home generation is not a viable way to cut electricity costs at this time. Equipment is reliable, but the capital outlays involved, the installation costs, and the operation and maintenance costs over the systems' life spans—probably 20 years or so—would usually make your cost per kwh higher than what you now pay to the electric company. The most expensive system—photovoltaics—may cost you 50¢ to $1 per kwh generated at current costs of equipment, assuming you had to borrow the money to purchase the system. Wind power is cheaper, with a range of as low as 11¢ per kwh up to 90¢ per kwh. Small-scale hydro power is the least expensive, running from 5¢ to 18¢ per kwh. There are places where high electric rates make home generation of electricity competitive, but generally speaking, this is not the case.

Things are changing, however. Overall system costs are coming down while utility rates keep climbing. Photovoltaic systems are expected to drop to a competitive level by the early 1990s. Small-scale wind power is also expected to drop in price while hydro power systems will be stable or may rise a bit. All these systems will be able to stand their ground against central generation of electricity, and they will very likely constitute a substantial portion of the residential electricity supply by the end of the first quarter of the next century.

Currently (and this may not change), the price of wind electricity per kwh goes down as system size increases from small to moderate. Thus, it seems that—based on economies of scale—electrical generation by larger turbines on multiunit wind farms is the ideal way to go. Solar generation with photovoltaics, on the other hand, is not affected by economies of scale—small systems can be just as economical as large systems—so it is likely that this is where the revolution in home electrical generation will occur. Access to small-scale hydro power is so limited that it will never constitute a major component of residential generation. Few people have suffiently powerful streams on their land to harness the water for electrical generation.

The technology for all three systems is quite reliable.

[1] Photovoltaic solar panels are mounted on a south-facing structure—usually a roof—and will convert the sun's energy directly into electricity. Most of the panels sold today produce around 35 watts of power at noon on a sunny day. Adjusting for (a) the varying lengths of daylight throughout the year, (b) the varying amounts of sunlight received at specific locations (the Southwest is best, the Northwest is worst), and (c) the orientation and tilt angle of the collectors, you can expect to average between 175 and 250 watts per day from each panel installed. An energy-conscious family using about 300 kwh a month would require 40 to 60 panels to meet all power needs.

The number of panels needed will drop as the efficiency of photovoltaic systems increases. A tripling of efficiency is certainly possible and would mean that 15 or 20 panels could meet a 300-kwh load. Technological breakthroughs in materials and design are also anticipated, and in conjunction with mass production and standardization, one can expect drastic price reductions on solar panels.

Bear in mind that besides buying the panels themselves, there are also certain "balance-of-system" costs involved—the price of the other equipment needed in the system. Balance-of-system costs are not likely to come down the way the cost of solar panels will, but overall system costs will nonetheless reach competitive levels within the coming decade. One balance-of-system component you will probably need is a power inverter. Since photovoltaic panels generate direct current (DC) and household appliances run on alternating current (AC), we must either convert to DC appliances or install a power inverter to convert DC electricity to AC. The latter is usually preferable.

Another factor to consider when tabulating balance-of-system costs arises from the need to coordinate household electric demand with the home's production of electricity. Maximum solar generation of power occurs at midday, and there is no generation at night. We can store excess power in batteries—thus raising our balance-of-system costs by the price of the batteries—or we can sell it to the utility company. If we follow the latter course, we will draw on utility power when our demand exceeds our system's output. The Public Utility Regulatory Policies Act of 1978 (PURPA) has made it easier to sell home-generated power to utilities, and it actually requires that your utility company pay you a "reasonable rate" for any surplus power you generate.

2 Wind-energy systems are subject to the same basic constraints (i.e., conversion of DC to AC; battery storage or utility-connection costs; and installation costs, including the price of a tower on which you will install your wind-powered electric generator). Even with anticipated technological advances, mass production, standardization, and the increased efficiency of wind turbines, do not expect anywhere near as great a drop in wind-power equipment costs as will occur with solar electricity. However, since costs for wind-power systems are already less than those for photovoltaics, the era of competitive wind power is close, particularly in areas with high average wind speeds (the Northeast, parts of the Midwest, northwestern coastal areas, and scattered mountain and valley locations).

3 Streams have a distinct advantage for electricity generation: They flow continuously, at least for most of the year. Unlike the sun and wind, the energy in a stream is virtually constant. The power may be small when measured over the short run, but it accumulates dramatically over the long run. The amount of water in the stream (its volume), how fast it flows, and its dropping distance (its "head") determine the stream's available power. Generally it is believed that a stream that produces less than 500 watts output is probably not worth developing for power generation.

All three systems—especially wind and hydro power—require extensive, professional site evaluations prior to a commitment to purchase and install equipment. Virtually any area of the country is all right for solar electric generation, although less sunny areas will require more panels than sunny locations. Sites with average wind speeds of 10 to 12 mph are generally marginal or not worth developing for wind power. Areas with 13- to 14-mph winds are good sites; those with 15 mph or more are excellent. Although these may seem like small differences, a 15-mph wind has twice the power potential of a 12-mph wind, while below 12 mph, many wind turbines will not even rotate. At potential water-power sites, the greater the drop over a given distance, the more likely that installing a small hydro system will be worth your while.

Photovoltaic Systems

Consumer Panel
Energy Sciences, Inc., Div. of Solarex
16730 Oakmont Ave.
Gaitherburg, MD 20877
(30-watt peak output.)

Photovoltaic Modules Genesis, M63, M73
ARCO Solar, Inc.
9351 Deering St.
P.O. Box 4400
Chatsworth, CA 91311
(5-, 30-, 40-watt peak output.)

Wind-Energy Systems

Bergey Excel
Bergey Windpower Co., Inc.
2001 Priestley Ave.
Norman, OK 73069
(10-kilowatt system.)

Whirlwind Series 2000, 4000, 9000
Wind Generators
Whirlwind Power Co.
207½ E. Superior St.
Duluth, MN 55802
(2-, 4-, and 9-kilowatt systems.)

Hydroelectric Systems

Mini-Tube Small Turbine
Allis-Chalmers Co.
P.O. Box 712
York, PA 17405
(Output would exceed power demand of most households and should be used with utility connection. Other small turbines also available.)

Index

Page numbers in italic indicate illustrations; page numbers followed by t indicate tables.

A Absorber plates, 78
Active-solar heating systems, 73
Active-solar water systems, 22–23, *23*, 24, 96
Aerator flow restrictors, 14, 103
AFUE (Annual Fuel Utilization Efficiency), 83, 127
Air circulation, 52–55, 122–23
Air conditioners
  best temperature for, 46
  Btu capacity of, 44, 119–20
  central, 46, 48, 120, 121
  conservation tips for, 56
  as dehumidifiers, 46
  electricity used by, 1, 4
  Energy Efficiency Rating (EER) of, 44, 120, 126
  energy-efficient, 92, 119–21
  energy labels on, 8
  misting devices for, 46–47, *47*, 120
  reverse-cycle, 48–49, 121
  room, 44, 46, 48, 120–21
Air conditioning
  benefits of fans for, 52–55, 122
  boosting efficiency of, 45, 48, 120
  clock thermostat for, 120
  electric space heating and, 48–49
  financial impact of conservation efforts for, 95t
  heat pumps and, 68, 126
  insulation and, 45, 45t, 50, 53, 55, 120
  timing for, 46, 115
  ventilation and, 123
  zoned, 44, 46, 120–21
Air leaks, 62
Air pollution, indoor, 93, 135
Air-to-air heat exchangers, 92–93, 135–36
Air-to-air heat pumps, 125–26
Air-to-water heat pumps, 102
All-electric homes, 6
Annual Fuel Utilization Efficiency (AFUE), 83, 127

Antifreeze for solar water heaters, 108
Appliances
    cutting costs of, 86–93
    electricity used by, 1, 4, 5t, 90–91
    energy labels on, 8
    energy-saving cooking, 39–40, 43
    energy-saving modifications to, 97–98
    financial impact of conservation efforts for, 95t
    life expectancy of, 92, 93t
    microelectronic control over, 116
    replacement of, 4, 92–93
    timing devices for, 115
Attached greenhouses, 75–77, 76, 84, 122–23, 130, 131–32
Attic fans, 52–53, 123–24
Automatic misting devices for air conditioners, 46–47, 47, 120
Automatic thermostats, 64, 84, 116

B Backup systems, 23, 81, 84
Baseboard electric-resistance heating systems, 125
Baseboard resistance space heaters, 1
Basement, insulation of, 60
Batch solar preheaters, 24
Batch water heaters, 21, 22, 109

Bathing, reducing water use during, 16
Batteries, power storage in, 96
Box fans, 124
Breadbox heaters, 109
British thermal unit (Btu), 5, 8, 44, 80, 119–20

C Carpeting, R-value of, 60
Catalytic combustors, 134
Catalytic wood stoves, 81, 82, 133, 134
Caulk, 45, 57, 61–63, 63
Ceiling, insulation above, 58–59
Ceiling fans, 52, 123, 124
Cellulose, insulating with, 59
Central air conditioners, 46, 48, 120, 121
Chimney-venting minifurnaces, 70, 71, 73
Circulation, air, 52–55, 122–23
Clock thermostats, 64, 84, 116
    for air conditioning, 120
Clothes dryer(s), 86–87
    converting, 87, 92
    cutting cost of, 86–87, 95t
    electronic ignition for, 87
    energy-efficient, 137
    energy use by, 137
    heat reclaimer for, 87
    solar, 86
Clothesline, benefits of, 86
Clothes washer(s)
    cutting cost of, 88
    energy-efficient, 137

energy labels on, 8
energy use by, 136, 137
reducing water use in,
16, 88
Coal, 6t, 82
Coal furnaces, 81–83
Coal stoves, 81–83
Coefficient of Performance
(COP) rating, 103, 125
Collectors, flat-plate, 107,
130
Compression gaskets, 62
Concrete slab, insulating
home built on, 60
Condensing furnaces, 128
Conservation
with dishwashers, 136
of electricity, 10, 11t, 96
heating and, 56–66
hot water and, 12–15
by renters, 97–98
Conservation retrofit
for cooling, 45
financial impact of, 95t
for heating, 66, 67t,
84–85
for hot water systems,
12–15, 24
minifurnaces and, 72
power use reduction due
to, 94, 95t
for refrigerators, 33–34
for space heating, 68
for windows, 61
Convection, natural, cooling
and, 52
Convection ovens, 40, 119
Cookers, solar, 42–43, *42*
Cooking
cutting costs of, 37–43

financial impact of
conservation efforts
for, 95t
with gas, 40
microwave, 41
supplemental devices for,
39–40, 43
Cooling
air movement and,
52–55
alternatives for, 49–55
benefits of insulation for,
45, 45t, 50, 53, 55,
120
cutting costs of, 44–56
fans and, 52–55, 122
home's capacity for, 49
humidity and, 52
interior heat sources and,
55
setbacks for, 116
timing devices for, 46,
115
vegetation and, 45,
51–52, 55–56
wind turbines and, 52
zoned, 44, 46, 120–21
COP (Coefficient of
Performance) rating, 103,
125
Cord of wood, 80–81

D Dehumidifiers, 46, 89
Derating, furnace efficiency
and, 127
Destratifiers, 64, *65*
Dimmer switches, installation
of, 29
Direct-gain systems, 73–75,
74t, 77, 84, 130–31

Direct-venting minifurnaces, 70, 71, *71*, 73, 128
Dishwashers, 87–88
    conservation with, 136
    energy-efficient, 137
    energy labels on, 8
    energy used by, 87, 136–37
    financial impact of conservation efforts for, 95t
    reducing water used in, 16, 88
Doors, 62, 64–66
Drapes, insulating, 61
Dryers, clothes, 86–87, 92, 95t, 137
Drying racks, indoor, benefits of, 87

E EER (Energy Efficiency Rating) for air conditioners, 44, 120, 126
Electrical energy, measurements of, 8
Electrical outlets, air infiltration through, 62
Electrical resistance heat, 23
Electric appliances, replacement of, 92–93
Electric baseboard systems, 84
Electric bill(s)
    checking excessive, 98–100
    economic stress due to, 1–2
    how to combat, 3–6
    for unoccupied house, 100–101

Electric company, 96–97, 139
Electric heat, 58, 66, 67t
Electric hot water systems, 12, 19
Electricity
    air conditioning and, 1, 4
    alternatives to, 6
    appliances and, 1, 4, 5t, 90–91
    assessing use of, 1–11
    cost of, 1, *2*, 6
    energy content of, per dollar, 6t
    heating with, 58, 66, 67t, 93, 125, 135
    home generation of, 96, 137–40
    for lighting, 25, 28–30, 31t, 32t
    saving, 10, 11t, 96
    selling, to power company, 96–97, 139
    for water heaters, 1, 4, 12, 15
Electric meter, how to read, 7, *7*
Electric ovens, 38, 119
Electric radiant heaters for task heating, 69–70
Electric range(s)
    costs of, 1, 37–38
    energy-efficient, 41, 118
    manufacturers of, 119
Electric rates, *2*
    factors affecting, 2–3, 9
    fluctuation in, 99
    off-peak, 8–10, 15
Electric-resistance heaters, 127

Electromagnetic energy,
    cooking with, 41
Electronic ignition, 87, 103
Energy
    conservation of, by
        renters, 97–98
    electrical, 8
    electromagnetic, 41
    renewable sources of, 96
    terms concerning, 8
Energy audits, 94
Energy devices, renewable, 4
Energy Efficiency Rating
    (EER) of air conditioners,
    44, 120, 126
Energy-efficient products,
    evaluation of, 94
Energy labels on appliances,
    8
Extruded polystyrene, 59

F Fans, 122–24
        attic, 52–53, 123–24
        ceiling, 52, 123, 124
        cooling and, 52–55, 122
        efficiency of, 55, 122
        functions of, 122–24
        for greenhouses, 131–32
        kitchen, heat loss
            through, 62
        manufacturers of, 124
        oscillating, 124
        portable box, 124
        for solar heating systems,
            80
        thermostatically
            controlled, 123
        variable-speed, 123
        wall, 123
        whole-house, 53–55,
            54, 124

        window, 55, 124
Faucets
    fixing leaky, 16
    flow restrictors for, 14, 103
    replacing washers in, 16
Fiberglass, insulating ability
    of, 14, 59
Film, window, 51, 111, 112,
    114
Fireplaces, heat loss through,
    62–63
Flat-plate collectors, 107, 130
Floors, insulation under, 60
Flow restrictors, 14–15, 24,
    103, 104
Flue heat-recovery, 127
Fluorescent light, 27–28, 27t
Freezers, 8, 36–37, 117–18
Frost-free refrigeration, 33
Fuel
    energy content of, 6t
    slowing furnace use of, 127
Furnaces
    coal, 81–83
    condensing, 128
    corrosion in, 128
    derating, 127
    efficiency of, 83, 127–29
    gas, 71, 83–84, 127
    manufacturers of, 129
    noncondensing, 128
    oil, 83–84, 128
    oversizing, 128
    retrofitting, 127–28
    slowing fuel consumption
        of, 127

G Gas, 6t, 17–18, 40, 83
    Gas furnaces, 71, 83–84, 127

Gas hot water systems, 24, 103
Gaskets
　compression, 62
　freezer, 36
　oven, 38
Gas ovens, 119
Gas ranges, 40, 43, 118–19
Gas water heater(s), 17, 24, 103, 104
　conversion to, 136
　tankless, 18–19, 24, 105–7
Glazing
　in direct-gain systems, 130–31
　greenhouse, 131
　heat loss through, 61
　layers of, solar heating and, 111
　low-iron, 111
　solar, 75, 130
Greenhouses, attached passive-solar, 75–77, 76, 84, 122–23, 130, 131–32

H Heat
　counteracting sources of, 119
　distribution of, minifurnaces and, 72
　interior, cooling and, 55
　measurements of, 8
　recovering, from furnace flue, 127
　redistributing, 64
　stratification of, 64
Heaters
　breadbox, 109
　chimney-venting, 70, 71, 73

direct-venting, 70, 71, 71, 73
electric radiant, 69–70
electric-resistance, 127
nonventing, 70, 128
portable kerosene, 68–69
quartz, 69
water. See Water heater(s)
window, 78, 79, 80, 130, 132
Heat exchangers, 92–93, 135–36
Heat flow, improving, 122
Heating
　caulk and, 57, 61–63
　conservation and, 56–66
　costs of, 57t, 72, 85t
　cutting costs of, 56–86
　furnaces for, 127
　infiltration and, 61
　insulation and, 57
　orientation of house and, 57
　savings possible with, 85–86
　space. See Space heating
　task, 69–70
　water. See Water; Water heater(s)
　weather stripping and, 45, 57, 61–63
　windbreaks and, 57
　with wood, 133–35
Heating coils, reducing use of, in appliances, 86–93
Heating system(s)
　active-solar, 73

Annual Fuel Utilization
Efficiency (AFUE)
rating for, 83, 127
conservation retrofit for,
66, 67t, 84–85
efficiency of, 6, 83, 127
electric, 58, 66, 67t, 93,
125, 135
supplemental, 66–84
timing devices for, 115
zoned, 68, 69
Heat load, 8, 77
Heat loss, 8, 61, 110
Heat pump(s), 66, 68
add-on, 126
advantage of, 68
air conditioning and, 68,
126
air-to-air, 125–26
air-to-water, 102
as backup system, 84
central heating with, 68
chimney-vented
minifurnace vs., 71
Coefficient of
Performance (COP)
rating for, 103
costs of, 20
disadvantage of, 68
effect of outdoor
temperature on, 126
installation of, 19
manufacturers of, 126
moisture buildup in, 126
payback period for, 20,
24, 68
performance of, 19–20,
125
purchasing, 126
recovery time for, 103

reverse-cycle, 56, 68,
125
savings with, 24, 68, 103
types of, 66, 68, 103
zoned heating with, 68
Heat pump water heaters,
19–20, 102–3, 104
Heat reclaimers, 87
Heat stratification, 64, 123
Home wind electrical
generation systems, 97
Hot water heaters. *See* Water
heater(s)
Hot water pipes, insulating,
16
Hot water systems. *See*
Water; Water heater(s)
House
cooling load for, 49
orientation of, heating
and, 57
preparation of, before
vacation, 100–101
tightening up, 45
Humidifiers, alternatives to,
87, 89
Humidity, cooling and, 52
Hydroelectric power, 137,
138, 139–40
Hydroelectric systems, 97,
140

I Ignition, electronic, 87, 103,
106
Indoor air pollution, 93, 135
Indoor drying racks, 87
Infiltration, 61–62
Instant-on televisions, 88–89
Insulating doors, 62, 64–66

Insulating drapes, 61
Insulating shutters, 61, 111–12
Insulation
  of basement, 60
  blown-in loose-fill, 60
  above ceiling, 58–59
  concrete slab construction and, 60
  cooling and, 45, 45t, 50, 53, 55, 120
  cost of, 58–59
  for direct-gain systems, 130
  effectiveness of, 8
  under floors, 60
  heating and, 57
  of hot water systems, 14, 16
  materials for, 59, 61
  most important area for, 50
  recommended amount of, 59
  R-value of, 57
  savings from, 59
  of walls, 59–60
  window, 61
    movable, 73–74, 111, 113

K Kerosene, 6t, 68, 71, 73
Kerosene heaters, 68–69
Kilowatt-hours (kwh), 1, 8, 12
Kitchen fans, heat loss through, 62

L Light(ing), 25–28
  avoiding too much, 29–30
  ceiling, 29
  cutting costs of, 25–32
  electric, 25, 28–30, 31t, 32t
  financial impact of conservation efforts for, 95t
  microelectronic control over, 116
  natural, 28, 30
  reading, 29
  timing devices for, 30, 115
  typical use of, 31t, 32t
  varying needs for, 29
Light bulbs, 25–30, 26t, 27t
Light buttons, 30
Light dimmers, 29
Liquid propane, 6t, 40
Lumens, definition of, 25

M Magnetic induction cooktops, 41
Microwave cooking, 41
Mineral wool, 59
Minifurnaces, 70–73, *71*, 84, 128, 129
Misting devices for air conditioners, 46–47, *47*, 56, 120, 121
Movable window insulation, 73–74, 111, 113
Movable window sun control, 114

N Natural gas, 6t, 83
Natural light, 28, 30
Noncatalytic stoves, 134
Noncondensing furnaces, 128

Nonventing minifurnaces, 70, 128

Nuclear power, 3

O Off-peak electric rates, 8–10, 15

Oil, heating, 6t, 83

Oil furnaces, 83–84, 128

On-peak rates, definition of, 9

Oscillating fans, 124

Ovens, 118–19

    checking seals of, 38

    convection, 40, 119

    electric, 119

    gas, 119

    self-cleaning, 38

    stratification in, 40

Overhangs, shading with, 50–51

Overheating

    greenhouse, 122–23, 131

    solar heating and, 80

Overlighting, 29–30

P Passive-solar attached greenhouses, 75–77, *76*

Passive-solar breadboxes, 21, *22*

Passive-solar retrofit, 75

Passive-solar space heating, 130–32

Passive-solar systems, direct-gain, 73–75, 74t, 77, 84, 130–31

Payback period, definition of, 8

Perforated aluminum foil, 50, 53

Photovoltaics, solar, 96, 137–40

Piezoelectric ignition for gas water heaters, 106

Pilot light, turning off, 106

Pipe insulation kits, 16

Plants, cooling and, 45, 51–52, 55–56

Plastic film for windows, 111, 112

Plexiglas, as insulation, 61

Point-of-demand water heaters, 18–19, 24

    converting, to gas, 103

    for dishwashers, 87

    tankless, 105

Pollution, indoor air, 93, 135

Polyisocyanurate, 59

Polystyrene board, 60

Portable kerosene heaters, 68–69

Precooling misters for air conditioners, 46–47, *47*, 56, 120, 121

Q Quadpane windows, 111

Quartz heaters, 69

R Radiant heaters for task heating, 69–70

Radiation, microwave ovens and, 41

Ranges, 118–19

    electric, 1, 37–38, 41, 118

    gas, 40, 43, 118–19

Rate shock, 1–2, 4

Reading, best light for, 29

Reflective aluminum foil, 55

Reflector light bulbs, 27, 30

Refrigerators, 32–37, *34*, 35t, 95t, 117

Regulated power combustion in gas-fueled hot water systems, 103
Renters, energy conservation by, 97–98
Reverse-cycle air conditioners, 48–49, 121
Reverse-cycle heat pumps, 56, 68, 125
Roof windows, 112
Room air conditioners, 44, 46, 48, 120–21
Russian stoves, 134
R-value
    of carpeting, 60
    definition of, 8
    of doors, 64–65
    of insulating shutters, 111
    of insulation, 57
    of windows, 111

S   Service period for electric bill, 99
Shade walls, 51
Shading, cooling costs and, 45
Shading devices, 50–51
Shower heads, 14, 103
Shutters, insulating, 61, 111–12
Skylights, 30, 52, 110–14
Small appliances, operating costs of, 90–91
Solar collectors, 108, 130. See also Thermosiphoning air panel (TAP) collectors
Solar cookers, 42–43, 42
Solar dryer, 86
Solar glazing, 75, 130

Solar greenhouses, 75–77, 76, 84, 122–23, 130, 131–32
Solar heat, 55, 77, 111, 131
Solar heating systems, 73, 80, 130–32
Solar photovoltaics, 96, 137–40
Solar preheaters, 20–24
Solar screens, 51
Solar thermosiphoning air panel (TAP) collectors, 78–80, 79, 130, 132
Solar water heaters, 96, 107–10
Space heating
    air conditioning and, 48–49
    conservation retrofit for, 68
    electricity for, 1, 4, 83
    passive-solar, 130–32
    setbacks for, 116
Spark pilot-light ignition for gas water heaters, 106
Standby losses, 12, 103, 105
Storm doors, 64–66
Storm windows, 61, 110, 111, 114
Stoves, 81–83, 134
Stratification, heat, 64
    fans and, 123
    in ovens, 40
Sun control, 112, 114
Sun screens, 112
Sunspaces, 130, 131, 132

T   Tank jacket for water heaters, 14
Tankless water heating,

18–19, 24, 105–7
TAP collectors. *See*
    Thermosiphoning air panel
    (TAP) collectors
Task heating, 69–70
Television, 88–89, 95t
Temperature
    air, determining comfort
        level and, 50
    average summer, 49t
    of hot water, 12, *13*
    internal, of greenhouse,
        77
    outdoor, heat pumps
        and, 126
    for refrigerator, 33
Tempering tasks, 20–21
Tension strips for air leaks,
    62
Thermal chimneys, air
    movement and, 52
Thermal doors, R-value of, 65
Thermal mass, *76, 80*
Thermosiphoning air panel
    (TAP) collectors, 78–80,
    79, 130, 132
Thermosiphoning water
    heaters, 109
Thermostat(s)
    air conditioner, 46
    automatic, 64, 84, 116
    setting back, 63–64, 70,
        115
    water heater, 12, *13,* 20,
        116
Tight houses, air pollution in,
    93
Timing devices, 15, 46, 115,
    116
Transfer efficiency rating, 17

U Urea formaldehyde, insulating
        with, 59

V Vacation, electric bill during,
        100–101
Variable-speed fans, 123
Vegetation, cooling and, 45,
    51–52, 55–56
Ventilation, 52, 123–24, 128
Vents, air movement and, 52

W Wall fans, 123
Walls, insulation of, 59–60
Washer(s)
    clothes
        cutting cost of, 88
        energy labels on, 8
        energy use by, 136,
            137
        operating costs of,
            88
        reducing water use
            in, 16, 88
    faucet, replacement of,
        16
Water
    conservation of, 16, 88,
        92
    draining systems of, 100
    heating, 12, 87
    hot, 12–24, *13*
    preheating, 20–22
Water heater(s)
    batch, 21, *22,* 109
    conservation measures
        for, 12–15, 24
    costs of, 18–19, 103
    electric, 12, 19
    electricity used by, 1, 4,
        12, 15

Water heater(s)(*continued*)
  energy-efficient, 102–4
  energy labels on, 8
  financial impact of
    conservation efforts
    for, 95t
  flow restrictors for,
    14–15, 103
  freeze protection for,
    108, 109
  gas, 17, 24, 103, 104
    converting to,
      17–18, 136
    tankless, 18–19, 24,
      105–7
  heat pump, 19–20,
    102–3, 104
  hot water produced by,
    18
  insulation of, 14, 16
  point-of-demand, 18–19,
    24, 87
  reducing thermostat in,
    12, *13*, 20, 116
  solar, 96, 107–10
    square footage for,
      109
  tankless, 18–19, 24,
    105–7
  thermosiphoning, cost of,
    109
  thermostat in, 12, *13*, 20
  timer or manual switch
    for, 15
Water pumps, electric, 92
Watt(s), 8, 25
Weather stripping, heating

and, 45, 57, 61–63
Whole-house fans, 53–55,
  54, 124
Windbreaks, heating and, 57
Wind electrical generation
  systems, home, 97
Window fans, 55, 124
Window film, 51, 111, 112,
  114
Window heaters, 78, *79,* 80,
  130, 132
Windows, 61, 110–14
  energy-conserving, 113
  heat lost through, 61,
    110
  insulation for, 61, 73–74,
    111, 113
  movable sun control for,
    114
  multiple-pane, 110–11
  plastic film for, 111, 112
  quadpane, 111
  roof, 112
  R-values of, 111
  solar heat from, 73, 111
  storm, 61, 110, 111, 114
Wind power systems, 137,
  138, 139–40
Wind turbine, cooling and, 52
Wood, 6t, 80–81, 133–35
Wood stoves, 80–81, *82,* 84,
  133–35

Z Zoned cooling, 44, 46,
    120–21
  Zoned heating, 68, 69